Introduction to Fluid-Structure Interactions

Yahya Modarres-Sadeghi

Introduction to Fluid-Structure Interactions

Springer

Yahya Modarres-Sadeghi
Department of Mechanical and Industrial Engineering
University of Massachusetts
Amherst, MA, USA

ISBN 978-3-030-85882-7 ISBN 978-3-030-85884-1 (eBook)
https://doi.org/10.1007/978-3-030-85884-1

This Springer imprint is published by the registered company Springer Nature Switzerland AG
The registered company address is: Gewerbestrasse 11, 6330 Cham, Switzerland

Acknowledgments

I would like to express my profound gratitude to Michael P. Païdoussis, Michael. S. Triantafyllou, and Emmanuel de Langre for forming the foundation of my understanding of fluid-structure interactions (FSI). In the process of writing this book, I benefited from discussions with several of my colleagues and friends. I am grateful to Rémi Bourguet, for our inspiring discussions on different aspects of FSI in the past several years and for his focused feedbacks on this book. I have had productive conversations with Jason Dahl and Luca Caracoglia on many topics related to FSI. My former and current graduate students, Shruti Ladge, Anil Jain, Lakshmi-Narayanan Ramanujam, Banafsheh Seyed-Aghazadeh, Pariya Pourazarm. Gary Chang, Daniel Carlson, Todd Currier, Tyler Gurian, Suyue Han, Bridget Benner, Pieter Boersma, Umang Patel, Adrian Carleton, and Daniela Caraeni, contributed to different aspects of this book, from producing some of the results to going through the initial versions of the text, to giving feedback on the contents. Erica DeWitt's meticulous work on re-drawing and improving the figures throughout the book is much appreciated. This book is the result of teaching a course on fluid-structure interactions at the University of Massachusetts Amherst in the past several years, and the students who participated in the course over the years motivated me with their feedbacks and their enthusiasm. I am grateful to my editor, Michael Luby, and his team at Springer for their careful handling of this book from its first draft to its present form.

Contents

Chapter 1
Introduction

Fluid-structure interaction (FSI) occurs when a flexible or flexibly mounted structure (a rigid structure that is supported by springs) is in contact with flow, and the flow forces result in the movement of the structure. Then the movement of the structure causes the flow forces to change. This change in flow forces, in turn, results in changes in the structure's displacement (Fig. 1.1). In such systems, we cannot focus only on the structure and ignore the flow around it, or focus on the flow and ignore the structure in contact with it. The two of them must be considered together.

When we study fluid mechanics, we assume that the boundaries of the fluid in contact with a structure are fixed, and we focus on what the fluid does. When we study dynamics or vibrations, we assume flow forces are negligible: A structure oscillates or moves in vacuum. In both cases, if we can no longer make these assumptions, then we have an FSI system. For example, when on a windy day a tree bends and its leaves oscillate, or when a flag flaps, or a wind chime makes music, we cannot assume that the boundaries of the flow field are rigid, or we cannot assume that there is no flow around the structure. The garden hose problem is another example of an FSI system: leave a garden hose unattended when the valve is wide open: the hose starts moving around. A hose made of steel would not have moved, neither a rubber hose with no water. When we snore, our soft palate oscillates because of the flow of air around it. We would not have snored if we did not breathe, or if we did not have a soft palate.

FSI is a growing field of research. Traditional observations of FSI phenomena have been in aerospace (flutter of an airplane wing), civil engineering (wind-induced oscillations of tall buildings and bridges – such as the very well-known example of the Tacoma Narrows Bridge's collapse), and ocean engineering (oscillations of offshore platforms due to waves). An offshore floating wind turbine is an example of several types of FSI systems: the waves interact with the platform and the platform moves. The currents are in contact with the mooring lines—the long cables that attach the platform to the seabed—and make them oscillate. The wind turbine tower interacts with the wind. If the blades are long, which is normally the case in a floating wind turbine blade, they bend as they rotate, and they oscillate due to the wind,

Y. Modarres-Sadeghi, *Introduction to Fluid-Structure Interactions*,
https://doi.org/10.1007/978-3-030-85884-1_1

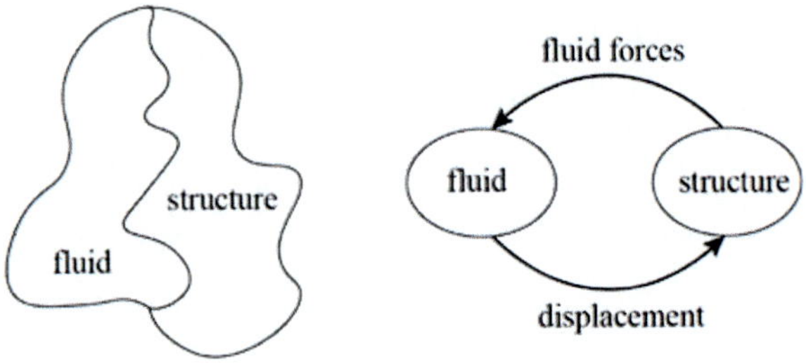

Fig. 1.1 A fluid-structure interaction system

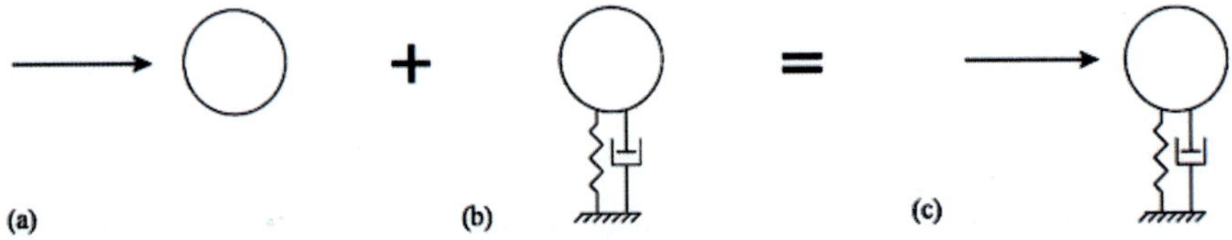

Fig. 1.2 (**a**) Flow around a fixed cylinder combined with (**b**) a mass-spring-damper constitutes a (**c**) fundamental FSI problem

because they are large, and therefore flexible. Several biomedical and biomechanics problems include FSI systems. Arteries are flexible structures that are constantly in contact with blood flow. A heart valve and a cerebral aneurysm are also flexible structures in contact with blood. We can find several other FSI examples around us. More researchers from different fields are becoming interested in FSI phenomena as they find more and more FSI examples in their work.

In fluid mechanics, a typical problem is the flow behavior around a rigid cylinder (Fig. 1.2a). What is the pressure distribution around the cylinder? At what angle will the flow separate from the cylinder? What is the thickness of the boundary layer? Questions of this type are answered in studying this problem. In dynamics and vibrations, a mass (with any cross section) supported by a spring and a damper is a fundamental problem (Fig. 1.2b). What is the natural frequency of the system? How does it respond to an external force? For how long does it oscillate after an initial disturbance? Questions of this type are typical questions that are answered when we study this problem in a dynamics or vibrations context. The combination of these two problems creates a fundamental problem in FSI: What will happen if we place a cylinder supported by a spring and a damper in the flow of some fluid (Fig. 1.2c)? How will the pressure distribution around the cylinder and as a result the flow forces acting on it change if the cylinder moves? Under which conditions will the cylinder move? How will its oscillations change compared with the case where it oscillates in vacuum? This example is the first FSI example that we will study in this book.

This book will start with a review of the fundamentals in fluid mechanics (Chap. 2) and dynamics and vibrations (Chap. 3) that we will need to relate to the concepts that we will discuss in the later chapters. Then we will start focusing on FSI systems from Chap. 4, where we will consider the case of a flexibly mounted

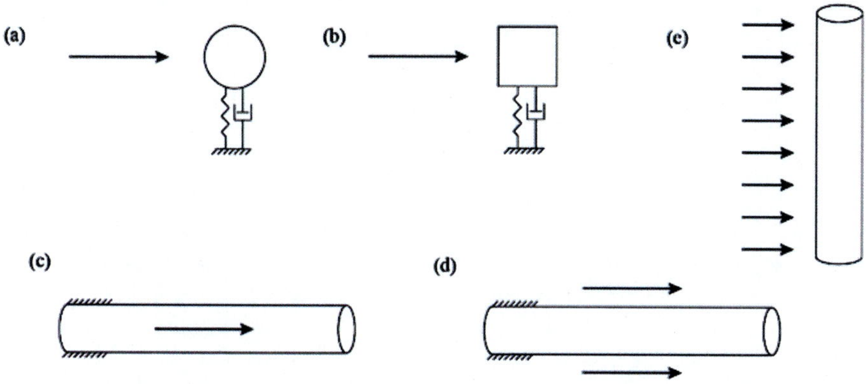

Fig. 1.3 A summary of the fundamental problems that we will discuss in this book: (**a**) a flexibly mounted circular body in flow, (**b**) a flexibly mounted noncircular body in flow, (**c**) a flexible pipe conveying fluid, (**d**) a flexible cylinder in axial flow, and (**e**) a flexible cylinder placed perpendicular to flow

cylinder placed in flow (Fig. 1.3a) and we will study vortex-induced vibration (VIV). VIV is a fundamental problem in FSI, and there are several review articles and book chapters on different aspects of VIV. Then in Chap. 5, we will extend our discussion on the response of a flexibly mounted structure placed in flow to structures with noncircular cross sections, such as a structure with a square cross section (Fig. 1.3b) or an airfoil supported by a spring and a damper and placed in flow. By considering these cases, we will discuss galloping and flutter phenomena, which are other fundamental FSI problems.

The concepts that we will discuss in Chaps. 4 and 5 are based on systems with one or two degrees of freedom. In these systems, the equations of motion of the structure are written in terms of one or two ordinary differential equations (ODEs). The structure is in contact with flow, whose behavior follows the Navier-Stokes equations, which are a set of nonlinear partial differential equations. In this book, we will not consider numerical solutions of Navier-Stokes equations using computational fluid dynamics (CFD) methods. Instead, we will use either experimental data to observe some phenomena or phenomenological models in which instead of Navier-Stokes equations, we will represent the fluid forces by equations that are based on our understanding of that particular system. These representations of the fluid forces will be equations that are functions of the structure's motion, since in any FSI system, the flow forces change as the structure moves. Then the fluid forces will also be represented by the same unknowns that represent the structure, and therefore the equations that will represent the FSI system will be a set of one or two ODEs as well.

After discussing the 1 DOF and 2 DOF systems, in Chap. 6 we will start focusing on flexible continuous structures, and specifically we will consider a flexible beam in contact with flow. In this system, the structure's equation of motion is written in terms of one or two partial differential equations (PDEs), since its motion is a function of both time and space—a beam that oscillates, oscillates with different magnitudes along its length, and at each location along its length, the magnitude changes with time. In Chap. 6 we will introduce some fundamental concepts on the beam equation, and we will discuss a method called the Galerkin method to solve PDEs that represent continuous structures.

In Chaps. 7 and 8 we will introduce a class of FSI problems known as axial flow problems. In axial flow problems, a flexible structure is placed in parallel to the direction of flow. A flag that is flapping on a windy day, or the blood inside the artery are examples of axial flow problems. We will start discussing axial flow problems by focusing on the fundamental problem of a flexible pipe conveying fluid in Chap. 7 (Fig. 1.3c). While it is a relatively simple system, with a relatively simple equation of motion, a flexible pipe conveying fluid exhibits very complicated and diverse dynamics. Different types of structural instabilities, such as pitchfork bifurcation, Hopf bifurcation, and torus are observed in a system consisting of a pipe conveying fluid with some slight alterations (such as adding extra supports or masses to the system). As a result of these different types of instabilities, different types of responses such as periodic, quasiperiodic, and chaotic are observed in these systems. Then in Chap. 8, we will study a case of an external axial flow problem in which the flow is parallel to the structure, and flows around the structure. We will consider a flexible beam with circular cross section (a flexible cylinder) in axial flow (Fig. 1.3d), and we will extend our understanding of axial flow problems to the cases where the flow is not limited to the inside of the structure. Axial flow problems have been used extensively to investigate different types of flow-induced instabilities and flow-induced responses that are observed in more complicated FSI systems, such as instabilities observed in wind turbine blades, airplane wings, or bridges.

In the last chapter of the book, we will continue discussing a flexible cylinder placed in flow, but this time the cylinder will be placed perpendicular to the direction of incoming flow (Fig. 1.3e), and as a result, the cylinder will undergo VIV. This will extend the VIV concepts that we will discuss in Chap. 4 for a 1 DOF or 2 DOF system to a continuous system.

The goal of this book is to prepare the reader to study and understand research articles on various FSI phenomena by providing important fundamental understandings of a variety of canonical problems in FSI. The aim is not to give a comprehensive review of the literature, but rather to go over the fundamentals that we will need to understand FSI phenomena. The book by Blevins (1990) reviews several FSI concepts and provides several references. The book by Païdoussis et al. (2014) provides a comprehensive review of crossflow FSI problems such as galloping and VIV., and Païdoussis's two-volume book (Païdoussis, 2014, 2016) is dedicated to axial-flow problems, both internal and external. The present book will prepare the reader to study and understand these more advanced books as well.

Chapter 2
Flow Around a Fixed Cylinder

1 Introduction

In this chapter, we focus on the fluid and discuss its behavior as it interacts with a fixed structure. We consider a body with a circular cross section (a cylinder), because circular cross sections are widely used in several applications, and also because a circular cross section is symmetric, independent from the direction of the incoming flow. A cylinder is an example of a bluff body. A bluff body is an object with a blunt cross section as opposed to a streamlined object. That means, for example, a circular, square, or elliptical cross section as opposed to an airfoil (placed in line with flow); or a plate placed perpendicular to the flow, as opposed to the same plate placed in parallel to the flow. Since the cylinder is a bluff body, many discussions in this chapter can be extended to structures with other bluff cross sections that are placed in flow. The presence of the bluff body changes the flow behavior from its behavior in a free stream and the flow exerts some forces on the body. Here we assume that the body is rigid and fixed, so that the flow forces cannot move or bend the body. This means that there is no fluid-structure interactions yet. The changes in flow pattern and forces from the flow on the structure are due to the presence of the structure only, not its motion. We discuss this case before we move on to the cases where the flow forces make the structure move. We will start those studies in Chap. 4. Zdravkovich (1997a, b) and Sumer and Fredsoe (2006) have discussed the flow around a cylinder in books. In this chapter, we focus on the concepts that we will need in the later chapters of this book.

Y. Modarres-Sadeghi, *Introduction to Fluid-Structure Interactions*,
https://doi.org/10.1007/978-3-030-85884-1_2

2 Separation and Vortex Shedding

If we place a cylinder in flow (Fig. 2.1), when the incoming flow sees the cylinder, the flow particles in contact with the cylinder tend to stop. This means that their velocities go toward zero, resulting in an increase in the local pressure. Assume the flow has a free stream pressure, P_0. The point where the flow sees the cylinder first and its velocity goes to zero is called the *stagnation point*. The pressure at the stagnation point is called the stagnation pressure, P_S, which is a very high pressure. This high pressure then pushes the flow particles to go around the cylinder. The flow particles directly in contact with the cylinder have zero (or very small) velocity, and the flow particles far from the cylinder have a constant velocity, equal to the velocity of the incoming flow, U. This implies that the flow velocity should vary from zero (where it is in contact with the cylinder surface) to the free stream velocity. This creates a layer of flow with a variable flow velocity, called the *boundary layer*. Let's think about the high stagnation pressure again. The particles are pushed to go around the cylinder, but there is no guarantee that they can make it to the other side of the cylinder, while they are still attached to the cylinder's surface. For high enough flow veloicties, at some point, rationally close to the widest section of the cylinder, the boundary layer cannot follow the cylinder's curvature anymore, and separates. When the boundary layer loses its contact with the cylinder's surface, we say *separation* has occurred. This is the case for boundary layers from each side of the cylinder: the upper side and the lower side. Then we will have two separated boundary layers (which we call *shear layers*) each from one side of the cylinder. If we consider the change in flow velocity within the boundary layer, the same variable flow velocity that existed in the boundary layer exists across the shear layer: the outer part has a higher flow velocity than the inner part. This implies that the shear

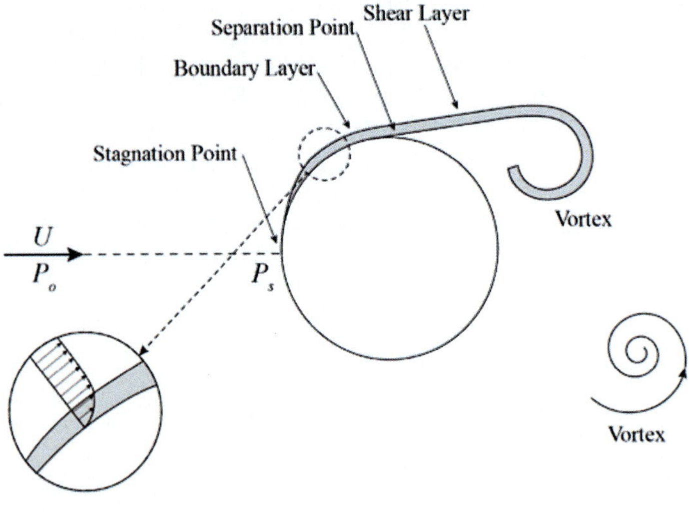

Fig. 2.1 Flow around a fixed cylinder

Fig. 2.2 A van Kármán Vortex Street behind a cylinder

layer bends inward and finally it "rolls" into the wake. We call the region behind the cylinder (after the separation point) and between the two shear layers the *cylinder's wake*. Then, in the wake, we have two shear layers that have rolled inward and can interact with each other. One rolled shear layer sees the other one and because they are rolling in opposite directions (since they came from two different sides) they can cut each other. When one shear layer cuts the other one, the latter shear layer makes a *vortex*, and we say this vortex is *shed*. The same steps are observed for the other shear layer, and we have a vortex that is shed from the other side. This process continues. We have to remember that the incoming flow is still coming, and another pair of vortices will be shed after the pair we just discussed. A global view of the wake shows that one vortex is shed from the top, and one from the bottom, again one from the top, and one from the bottom: this is called a *van Kármán Vortex Street* (Fig. 2.2).

3 Vortex Shedding versus Reynolds Number

An important question is how the velocity of the incoming flow influences the behavior we just discussed. Obviously, at the extreme case of a zero incoming flow velocity none of what we discussed can occur. But, what is the minimum flow velocity at which we do observe the separation, the shear layers, and the shed vortices? It is generally agreed that using dimensionless numbers is a better approach when we discuss a general behavior in a system. The appropriate dimensionless number for this discussion is the well-known *Reynolds number*, which is the ratio between the inertial forces and the viscous forces and is defined as:

$$Re = \frac{UD}{\nu} \equiv \frac{\text{Inertial Forces}}{\text{Viscous Forces}},$$

where U is the free stream flow velocity in m/s, D is the cylinder diameter in m, and ν is the kinematic viscosity of the fluid in m^2/s. The kinematic viscosity can be written as:

$$\nu = \frac{\mu}{\rho},$$

where μ is the dynamic viscosity of the fluid ($Pa.\ s$) and ρ is its density (kg/m^3). It is useful to have an idea about the values of the kinematic viscosity for typical fluids: water and air. For water, $\nu_{\text{water}} \sim 10^{-6}\ m^2/s$ and for air, $\nu_{\text{air}} \sim 1.5 \times 10^{-5}\ m^2/s$. They vary with the temperature a bit, but these numbers give us a good estimate. Note that the kinematic viscosity of air is one order of magnitude larger than that of water. With these values, we can quickly estimate the Reynolds number. This is one of the first questions that everyone asks us as soon as we talk about the flow.

After having an appropriate dimensionless number in hand, we can discuss how the Reynolds number affects the flow behavior around a cylinder.

At low Re ($Re < 5$), no separation is observed. This is the range for which the *potential flow assumption* works perfectly. We will discuss the potential flow assumption later.

For a range of $5 < Re < 47$, a symmetric pair of vortices is observed in the wake, which are fixed and do not flow downstream. These are called *Föppl vortices*. We see these in Fig. 2.3a based on some experiments at $Re = 31$. The vortices are right next to the cylinder and are not shed to the wake. The wake stays stable.

Starting at $Re = 47$, a behavior similar to what we discussed in Sect. 2 starts, and we observe vortices that are shed in the wake. This behavior persists for a wide range of Reynolds numbers with slight changes here and there: In the range of $47 < Re < 150$, *two-dimensional shedding* is observed, where two-dimensional implies that if we move along the length of the cylinder and take a snapshot of the flow around the cylinder at any cross section, we will get the same image every time. In this range, laminar vortices are shed from both sides of the cylinder and create a laminar vortex street (Fig. 2.3b). In the range of $150 < Re < 300$, transition to turbulence is observed in the vortex (Fig. 2.3c), which then leads to *3D shedding*, and fully turbulent vortex street for a very wide range of Reynolds numbers: $300 < Re < 3 \times 10^5$ (Fig. 2.3d). This range of Reynolds numbers is called the subcritical range. In the subcritical range, flow separation occurs at an angle of about 80°, when 0° corresponds to where the flow sees the cylinder first, i.e., the stagnation point (Fig. 2.4).

At higher Reynolds number range of $3 \times 10^5 < Re < 3.5 \times 10^6$, the boundary layer also becomes turbulent, resulting in a separation point at 140° (instead of 80°) and thus a narrower wake (Fig. 2.4). This is an important quantitative change in the behavior, since the new location for the separation point and the resulting narrower wake have significant influence on the flow forces acting on the cylinder as we will discuss later. In this range, the wake is disorganized and no vortex street is observed.

For larger Reynolds numbers ($Re > 3.5 \times 10^6$), the turbulent vortex street is re-established, while the wake stays narrower compared with the subcritical range. This range is called the *supercritical range*. The separation point stays close to 140° for this range as well. Lienhard (1966) was the first to summarize this dependency of the wake on the Reynolds number for a cylinder. This dependency is summarized in Table 2.1.

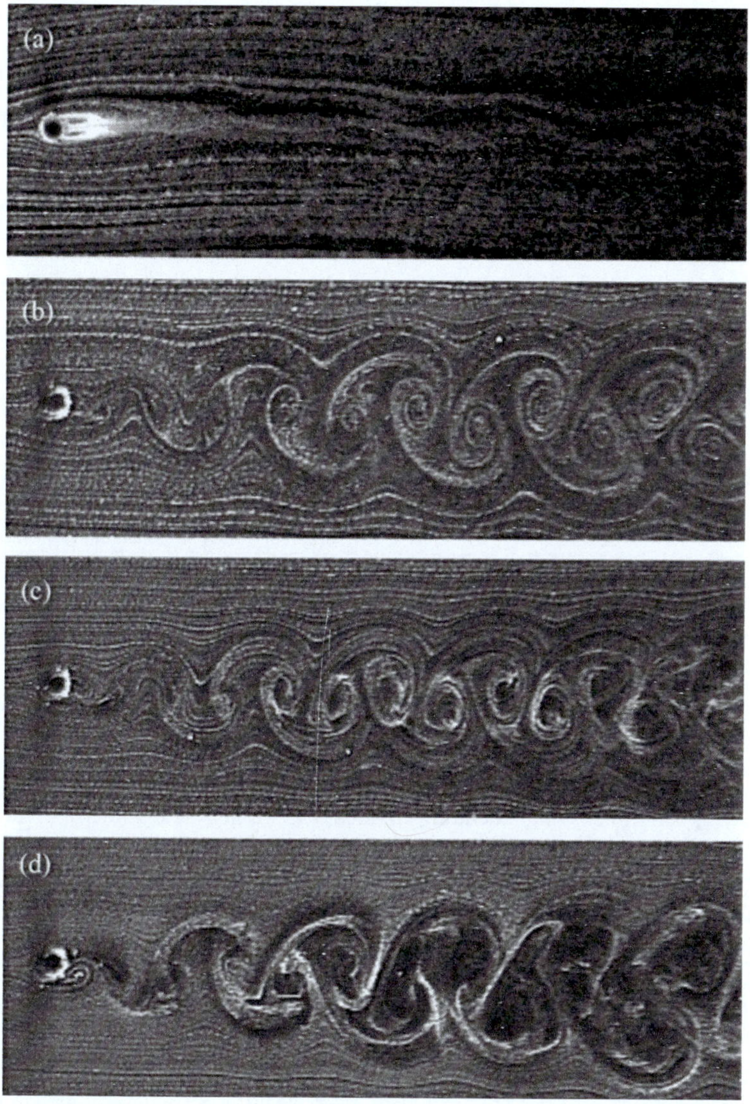

Fig. 2.3 Flow visualizations of the wake of a cylinder at (**a**) $Re = 31$, (**b**) $Re = 140$, (**c**) $Re = 190$, and (**d**) $Re = 330$

4 Strouhal Number

Another important question to ask when a cylinder is placed in flow is how often the vortices are shed. Or, what is the shedding frequency? It turns out that the frequency of vortex shedding depends on the incoming flow velocity, U, and the cylinder's

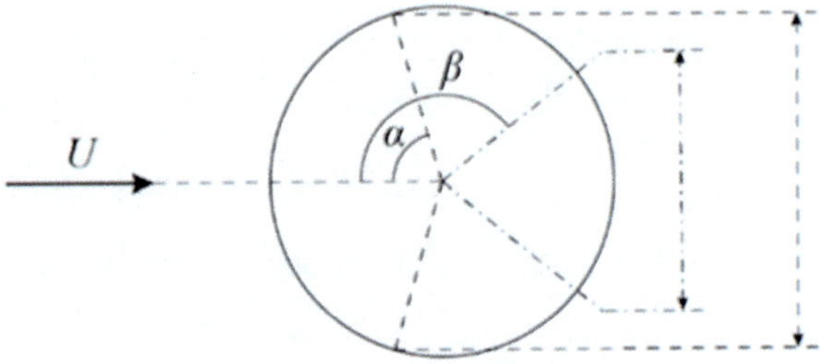

Fig. 2.4 Separation points and the width of the wake of a cylinder for $300 < Re < 3 \times 10^5$ (dashed line, $\alpha = 80°$) and $Re > 3 \times 10^5$ (dashed-dotted line, $\beta = 140°$)

Table 2.1 The dependency of the cylinder's wake on the Reynolds number

$Re < 5$	No separation
$5 < Re < 47$	A symmetric pair of vortices (Föppl vortices)
$47 < Re < 150$	Two-dimensional shedding
$150 < Re < 300$	Transition to turbulence occurs in the vortex
$300 < Re < 3 \times 10^5$	3D shedding, and fully turbulent vortex street
$3 \times 10^5 < \mathrm{Re} < 3.5 \times 10^6$	Disorganized and narrow wake
$\mathrm{Re} > 3.5 \times 10^6$	Turbulent vortex street and narrow wake

diameter, D, and can be represented through a dimensionless number, called the *Strouhal number*:

$$St = \frac{f_s D}{U}, \tag{2.1}$$

where f_s is the dimensional shedding frequency in Hz, D the cylinder's diameter in m, and U the flow velocity in m/s. The Strouhal number is a dimensionless frequency of vortex shedding, which remains almost constant for a wide range of Reynolds numbers. For $300 < Re < 300{,}000$, the Strouhal number is approximately equal to 0.2 for a cylinder. This is a very helpful fact. Imagine we have a cylinder with a given diameter and we plan on placing it in flow. We can use Eq. (2.1), with $St = 0.2$, and find the dimensional shedding frequency off the cylinder, without conducting any experiment or numerical simulation.

If we are out of that wide range of the Reynolds numbers, then the Strouhal number is not equal to 0.2 anymore. For $Re < 300$, the Strouhal number increases from 0.1 to 0.2 with increasing Reynolds number and for $Re > 300{,}000$, the Strouhal number increases from 0.2 with increasing Reynold number. Figure 2.5 shows how the Strouhal number varies with the Reynolds number. If the cylinder's surface is smooth, then the Strouhal number increases to much higher values for large Reynolds numbers. For a rough surface, the Strouhal number stays close to 0.2 for an even wider range of Reynolds numbers. A constant Strouhal number for a wide range of Reynolds numbers means that if we place a cylinder in flow, and increase

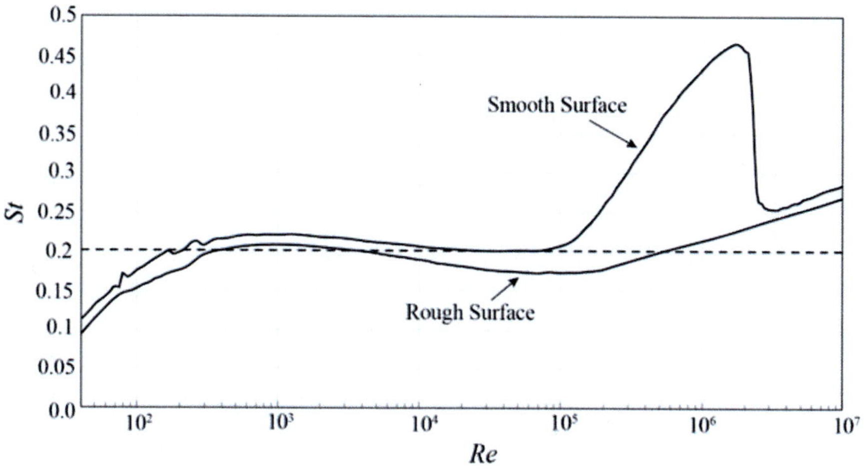

Fig. 2.5 Strouhal number versus Reynolds number for a cylinder. (Reproduced from MIT OCW Figure)

the flow velocity, then the shedding frequency increases linearly with flow velocity: $f_s = St\,U/D \equiv C\,U$, where C is a constant. Then we can say that the shedding frequency is following the *Strouhal law*. Later on in Chap. 4, we will see that the shedding frequency will not follow the Strouhal law, when the cylinder starts to oscillate. It is important to note here that the Strouhal number is reserved for the dimensionless shedding frequency when the cylinder, or any other bluff body that causes the shedding, is fixed. If the structure starts to move, then the dimensionless frequency of shedding observed in the wake of the structure should *not* be called the Strouhal number or the Strouhal frequency.

Vortex shedding is observed in the wake of any bluff body placed in flow. The body does not have to have a circular cross section. For any bluff body in flow, we will observe the flow separation, the interacting shear layers, and the vortex shedding. This means that we will have a frequency of vortex shedding, which we can non-dimensionalize, and call the Strouhal number for that specific bluff body. Obviously, the Strouhal number changes with the cross section of the bluff body. These Strouhal numbers are normally measured experimentally. For any given cross section, we will have to define a reference length, something equivalent to the cylinder diameter when we defined the Strouhal number for a cylinder. The Strouhal number for different geometries can be found in tables and graphs, and some of them are given in Fig. 2.6. In cases of a noncircular cross section, the orientation of the body with respect to the incoming flow can influence the Strouhal number too.

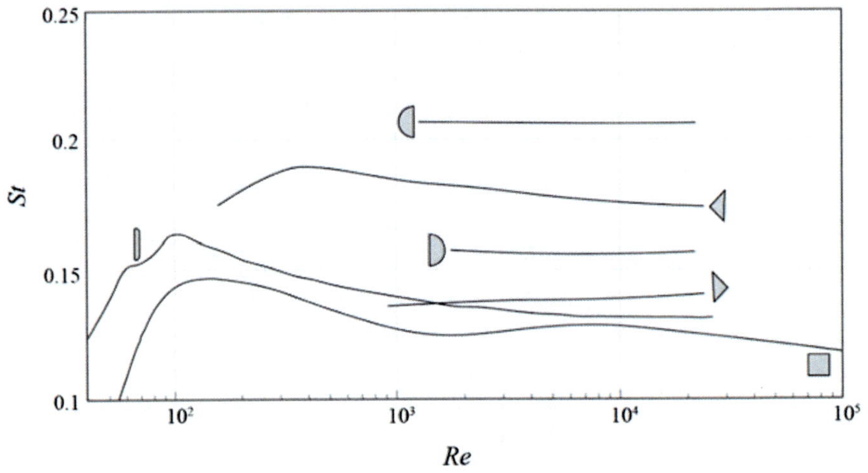

Fig. 2.6 Samples of Strouhal numbers for bluff bodies with different cross sections. (Reproduced from Blevins, 2001)

Fig. 2.7 The lift and drag forces acting on a cylinder placed in flow

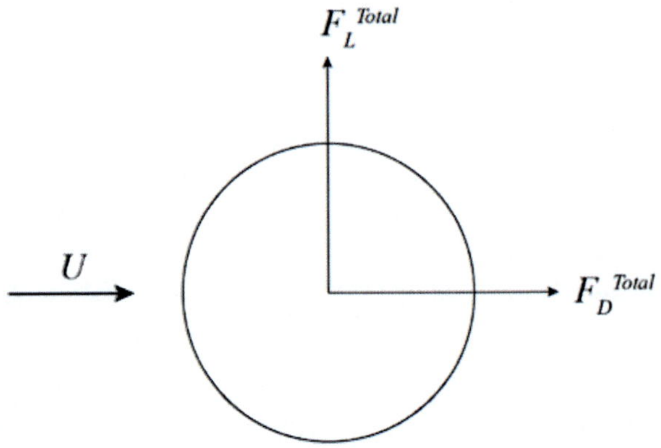

5 Flow Forces on a Fixed Cylinder

In this section, we focus on flow forces that act on a body with a circular cross section when it is placed in flow. Once the cylinder is placed in flow, flow forces act on it in both the transverse and inline directions (Fig. 2.7). We call the flow force that acts in the transverse direction, the *lift force*, F_L^{Total}, and the flow force that acts in the inline direction, the *drag force*, F_D^{Total}.

Both lift and drag forces have a mean and a fluctuating component. In what follows we will discuss these components separately. We start with the mean components, and use the potential flow assumption to discuss them. Through this discussion, we will see how the potential flow assumption works, and we will also see its limitations. Then we will discuss the fluctuating components of the flow forces—the components that are due to the vortex shedding off the cylinder—in both the crossflow and the inline directions.

5.1 *The Mean Flow Forces Acting on a Cylinder*

We will use the potential flow assumption to calculate the mean flow forces that act on a cylinder. The potential flow assumption is a popular assumption, since the resulting equations are very easy compared with the Navier-Stokes equations, which are the true representation of the flow forces. The potential flow assumption is valid for an inviscid, incompressible, and irrotational flow. Any introductory book on fluid mechanics (e.g., Fox & McDonald, 1998) has a discussion on the potential flow. Here, we will go over its basics and what we need for our future discussions in this book.

In potential flow, it is assumed that there is a velocity potential, ϕ, which has to satisfy the Laplace equation:

$$\nabla^2\phi = 0. \tag{2.2}$$

The Laplace equation in a two-dimensional case can be written as

$$\frac{\partial^2\phi}{\partial x^2} + \frac{\partial^2\phi}{\partial y^2} = 0, \tag{2.3}$$

and then the velocity components in the x- and y-directions can be found as

$$V_x = \frac{\partial\phi}{\partial x}, \tag{2.4}$$

and

$$V_y = \frac{\partial\phi}{\partial y}. \tag{2.5}$$

By assuming

$$x = r\cos\theta, \tag{2.6}$$

and

$$y = r\sin\theta, \tag{2.7}$$

the Laplace equation will be transformed into its cylindrical coordinates:

$$\frac{\partial^2 \phi}{\partial r^2} + \frac{1}{r}\frac{\partial \phi}{\partial r} + \frac{1}{r^2}\frac{\partial^2 \phi}{\partial \theta^2} = 0, \tag{2.8}$$

where the velocity components in the r and θ directions can be found as

$$V_r = \frac{\partial \phi}{\partial r}, \tag{2.9}$$

and

$$V_\theta = \frac{1}{r}\frac{\partial \phi}{\partial \theta}. \tag{2.10}$$

Then the flow around any body can be represented by a flow potential. The flow potential for a cylinder placed in uniform flow is

$$\phi = U\left(r + \frac{R^2}{r}\right)\cos\theta, \tag{2.11}$$

which will result in the following flow velocity components:

$$V_r = \frac{\partial \phi}{\partial r} = U\left(1 - \frac{R^2}{r^2}\right)\cos\theta, \tag{2.12}$$

and

$$V_\theta = \frac{1}{r}\frac{\partial \phi}{\partial \theta} = -U\left(1 + \frac{R^2}{r^2}\right)\sin\theta. \tag{2.13}$$

Once we have the flow velocity components, we can find the pressure distribution around a cylinder. First, Bernoulli's equation is valid, which means

$$P_0 + \frac{1}{2}\rho U^2 = P + \frac{1}{2}\rho V^2, \tag{2.14}$$

where P_0 and U are the freestream pressure and velocity, and P and V are the pressure and velocity on the surface of the cylinder. Then from Eq. (2.14), the relative pressure at the surface, p, is

$$p = P - P_0 = \frac{1}{2}\rho\left(U^2 - V^2\right). \tag{2.15}$$

We can find the velocity on the surface of the cylinder using its two components as:

$$V^2 = V_r^2 + V_\theta^2, \tag{2.16}$$

and then, using Eqs. (2.12) and (2.13), as:

$$V^2 = U^2\left(1 - \frac{2R^2}{r^2} + \frac{R^4}{r^4}\right)\cos^2\theta + U^2\left(1 + \frac{2R^2}{r^2} + \frac{R^4}{r^4}\right)\sin^2\theta, \tag{2.17}$$

which will be simplified to

$$V^2 = U^2\left(1 + \frac{2R^2}{r^2}(\sin^2\theta - \cos^2\theta) + \frac{R^4}{r^4}\right), \tag{2.18}$$

and using

$$\cos 2\theta = \cos^2\theta - \sin^2\theta, \tag{2.19}$$

it can be written as:

$$V^2 = U^2\left(1 + \frac{2R^2}{r^2}(-\cos 2\theta) + \frac{R^4}{r^4}\right), \tag{2.20}$$

which at the surface, for $r = R$, becomes:

$$V^2 = U^2(2 - 2\cos 2\theta). \tag{2.21}$$

Therefore, using Eq. (2.15), the relative pressure at the surface, p, will be

$$p = P - P_0 = \frac{1}{2}\rho(U^2 - U^2(2 - 2\cos 2\theta)), \tag{2.22}$$

or

$$p = P - P_0 = \frac{1}{2}\rho U^2(2\cos 2\theta - 1), \tag{2.23}$$

and since we know

$$\cos 2\theta = 1 - 2\sin^2\theta, \tag{2.24}$$

Equation (2.23) can be written as:

$$p = P - P_0 = \frac{1}{2}\rho U^2\left(1 - 4\sin^2\theta\right). \tag{2.25}$$

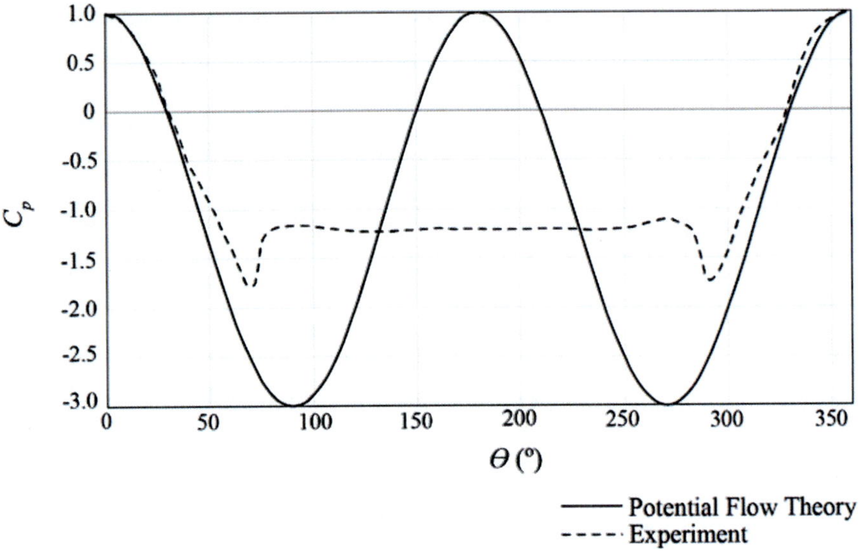

Fig. 2.8 Pressure coefficient around a cylinder based on experimental results and as predicted by the potential flow assumption. Figure reproduced from Sumer and Fredsoe (2006). In this plot, $\theta = 0°$ corresponds to where the flow sees the cylinder first

We can then define a dimensionless pressure coefficient as:

$$c_P = \frac{p}{\frac{1}{2}\rho U^2} = 1 - 4\sin^2\theta, \tag{2.26}$$

which depends on θ only and shows how the pressure is distributed on the cylinder's surface (Fig. 2.8).

Now that we have the pressure distribution (Eq. 2.25), we can integrate it around the cylinder to find the forces that act from the flow on the cylinder. The pressure distribution that we have found based on the potential flow assumption is symmetric with respect to a vertical plane that goes through the center of the cylinder. This means that any force that the cylinder feels on the right-hand side, it feels the same force on the left-hand side as well, but in the opposite direction. This implies that if we integrate the pressure over the entire cylinder, we will find a zero force in the inline direction, and as a result the cylinder does not feel any force from the flow in the direction of flow, i.e., no drag. This seems a bit strange to have a cylinder in flow, but have no force from the flow on the cylinder. This is called the *d'Alembert's paradox*. It is not difficult to guess the reason for the paradox, as you might have already guessed.

The pressure distribution, based on experimental results, is not symmetric with respect to the vertical plane. In fact in the upstream side of the cylinder ($\theta < 90°$ and $\theta > 270°$), the pressure follows the potential flow assumption more or less, but in its downstream side ($90° < \theta < 270°$), the pressure is almost constant, as opposed to what the potential flow assumption predicts (see Fig. 2.8). We can think about our discussion on the wake of the cylinder and guess the reason for this deviation from the potential flow assumption in the downstream side of the cylinder. The potential flow assumes that the flow has no viscosity. This means that there will be no boundary layer, and then no separation, and no shear layers, and no shedding. If there is no separation, then the pressure distribution will be symmetric with respect to a vertical line that goes through the center of the cylinder, and the drag force acting on the cylinder will be zero. This is true only for $Re < 5$, as we discussed in Sect. 2.3. For $Re > 5$, however, we do measure a nonzero value of drag, because separation does occur.

Now instead of assuming a pressure distribution following the potential flow assumption for all around the cylinder, we can assume that the pressure in the upstream half of the cylinder ($-90 < \theta < 90°$) follows the potential flow assumption (which looks like a fine assumption based on the results of Fig. 2.8) and in the downstream half ($90° < \theta < 270°$) is a constant with a pressure coefficient of $c_p = -1$, or

$$c_P = \begin{cases} 1 - 4\sin^2\theta & -90° < \theta < 90° \\ -1 & 90° < \theta < 270° \end{cases} \tag{2.27}$$

We can now calculate the mean drag force that is acting on the cylinder as:

$$\overline{F}_D = \int_0^{2\pi} p\cos\theta R d\theta, \tag{2.28}$$

where $\cos\theta$ is used to calculate the component in the inline direction.

Then:

$$\overline{F}_D = \frac{1}{2}\rho U^2 \int_0^{2\pi} c_P \cos\theta R d\theta, \tag{2.29}$$

and with the pressure distribution of Eq. (2.27):

$$\overline{F}_D = \frac{1}{2}\rho U^2 R\left[\int_0^{\frac{\pi}{2}} \left(1-4\sin^2\theta\right)\cos\theta d\theta + \int_{\frac{\pi}{2}}^{\frac{3\pi}{2}} (-1)\cos\theta d\theta + \int_{\frac{3\pi}{2}}^{2\pi} \left(1-4\sin^2\theta\right)\cos\theta d\theta\right], \tag{2.30}$$

which we can expand to:

$$\overline{F}_D = \frac{1}{2}\rho U^2 R\left[\int_0^{\frac{\pi}{2}} \cos\theta d\theta - 4\int_0^{\frac{\pi}{2}} \sin^2\theta\cos\theta d\theta - \int_{\frac{\pi}{2}}^{\frac{3\pi}{2}} \cos\theta d\theta + \int_{\frac{3\pi}{2}}^{2\pi} \cos\theta d\theta - 4\int_{\frac{3\pi}{2}}^{2\pi} \sin^2\theta\cos\theta d\theta\right], \tag{2.31}$$

and then integrate to find

$$\begin{aligned}\overline{F}_D &= \frac{1}{2}\rho U^2 R\left[\left(\sin\theta - \frac{4}{3}\sin^3\theta\right)\Big|_0^{\frac{\pi}{2}} - \sin\theta\Big|_{\frac{\pi}{2}}^{\frac{3\pi}{2}} + \left(\sin\theta - \frac{4}{3}\sin^3\theta\right)\Big|_{\frac{3\pi}{2}}^{2\pi}\right] \\ &= \frac{1}{2}\rho U^2 R\left[-\frac{1}{3} - (-2) + \left(-\frac{1}{3}\right)\right],\end{aligned} \tag{2.32}$$

which gives us a nonzero mean drag force:

$$\overline{F}_D = \frac{4}{3}\rho U^2 R. \tag{2.33}$$

Of course we find a nonzero value, because we just broke the symmetry of the pressure distribution that we had observed based on the potential flow assumption.

If we calculate the mean force in the vertical direction, i.e., the mean lift force,

$$\overline{F}_L = \int_0^{2\pi} p\sin\theta R d\theta, \tag{2.34}$$

however, we always find a zero value, using either the potential flow assumption or a constant pressure in the downstream side. The experimental results also show a zero mean force in that direction. This is expected, since everything is symmetric with respect to a horizontal plane that goes through the center of the cylinder, and that, of course, means a zero mean force in that direction.

Force is a dimensional quantity. In order to make the drag and lift forces dimensionless, we divide them by $\frac{1}{2}\rho D U^2$ to define the *mean drag coefficient* as:

$$\overline{C}_D = \frac{\overline{F}_D}{\frac{1}{2}\rho D U^2}, \tag{2.35}$$

and the *mean lift coefficient* as:

$$\overline{C}_L = \frac{\overline{F}_L}{\frac{1}{2}\rho D U^2}. \tag{2.36}$$

For a cylinder, $\overline{C}_D$ is nonzero, but $\overline{C}_L$ is zero.

5.2 *The Fluctuating Drag and Lift Forces*

In this section, we discuss the fluctuating forces that act on a cylinder due to the vortices that are shed in its wake. Due to the vortex shedding, the pressure

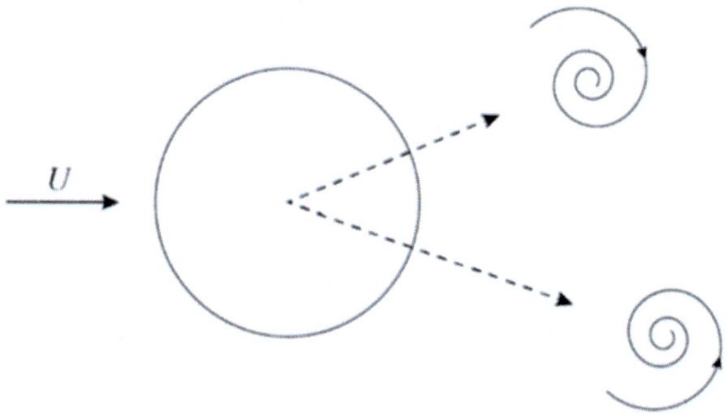

Fig. 2.9 Schematic of the locations of vortices shed in the wake of a cylinder, and the directions of forces that are exerted on the cylinder due to the shedding of each of the vortices

distribution around the cylinder changes periodically, resulting in a periodic force on the cylinder. We quantified the periodicity in the shedding by defining the dimensionless frequency of vortex shedding, the Strouhal number. We can think of every vortex that is shed from the cylinder as a point of low pressure. When the cylinder feels one of these points of low pressure, it wants to move toward that point. It cannot, because we have made sure that the cylinder is fixed and cannot move (later on we will discuss the cases where the cylinder is not fixed, but for now, let's assume that it is fixed and remains fixed). Each point of low pressure (each vortex) therefore exerts a force in an inclined orientation to the cylinder. Inclined because the vortex is shed from "north east" or "south east" of the cylinder (Fig. 2.9). Because of this orientation, we can find two components for this fluctuating force: one in the inline direction (*fluctuating drag force*) and one in the transverse direction (*fluctuating lift force*). Now let's think about the frequencies of these forces. In the transverse direction, the cylinder feels that an external force is acting on it with a frequency equal to the vortex shedding frequency: one vortex is shed from the top and one from the bottom in one period and the fluctuating lift force has the same frequency as the shedding frequency. But, what does the cylinder feel in the inline direction? One vortex from the top exerts a force with a component in the inline direction. The next vortex from the bottom exerts another force in the inline direction. Then in one periodic of shedding (one vortex shed from the top and one from the bottom) the cylinder feels two external forces in the inline direction—there is no difference between the inline component of the force due to the vortex shed from the top and the vortex shed from the bottom. This means that the fluctuating drag force has a frequency twice the shedding frequency.

Then we can write the fluctuating lift force with a frequency equal to the shedding frequency, $\omega_s = 2\pi f_s$, as

$$F_L(t) = \frac{1}{2}\rho D U^2 C_L' \sin(\omega_s t + \alpha), \tag{2.37}$$

and the fluctuating drag force, with a frequency twice the shedding frequency, $2\omega_s$, as

$$F_D(t) = \frac{1}{2}\rho D U^2 C'_D \sin(2\omega_s t + \beta), \tag{2.38}$$

where α and β are the phase differences with respect to the vortex shedding, ω_s is the shedding frequency, and C'_L and C'_D are constants proportional to the magnitudes of the fluctuating lift and fluctuating drag forces, respectively. The phase differences quantify the delay that exists between the time of the shedding of a vortex and the time that the cylinder feels the force due to that vortex. This delay does not have to be equal in the lift and drag directions and therefore α and β are not necessarily equal.

We can define new coefficients for the fluctuating lift and drag forces, C_L and C_D, which also include the frequency of the fluctuation as

$$C_L = C'_L \sin(\omega_s t + \alpha), \tag{2.39}$$

and

$$C_D = C'_D \sin(2\omega_s t + \beta). \tag{2.40}$$

Then, the fluctuating lift and drag forces can be written in terms of the fluctuating lift and drag coefficients as:

$$F_L(t) = \frac{1}{2}\rho D U^2 C_L, \tag{2.41}$$

and

$$F_D(t) = \frac{1}{2}\rho D U^2 C_D. \tag{2.42}$$

The total lift and drag forces that act on the cylinder include both the mean and fluctuating components:

$$F_L^{Total} = \overline{F_L} + F_L(t), \tag{2.43}$$

and

$$F_D^{Total} = \overline{F_D} + F_D(t). \tag{2.44}$$

The mean value of the lift for a cylinder is equal to zero ($\overline{F_L} = 0$), due to the symmetry of the structure, but this is not necessarily the case for other bluff bodies, since they are not necessarily symmetric.

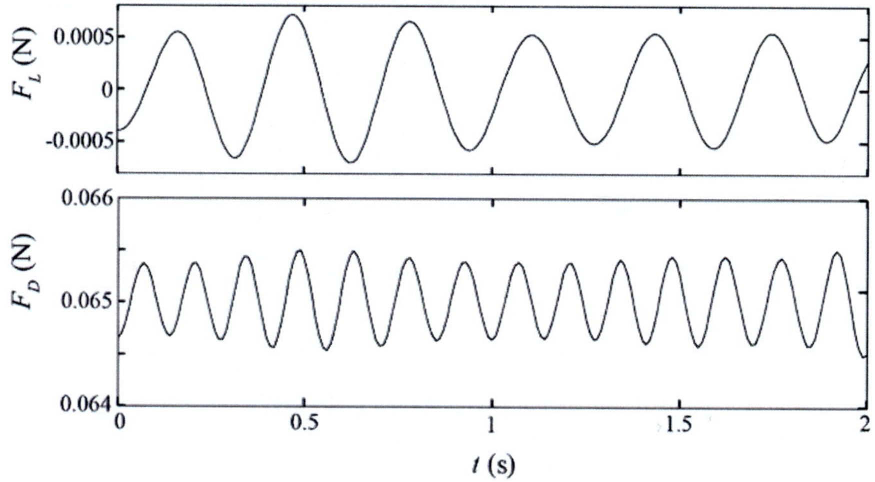

Fig. 2.10 Time histories of lift and drag forces acting on a cylinder at a Reynolds number of $Re = 2900$ based on experimental measurements

Figure 2.10 shows sample experimental measurements of the lift and drag forces that act on a cylinder placed in flow. The mean value of the lift force is zero, as expected, but the mean value of the drag force is not. The frequency of the drag force is twice the frequency of the lift force.

6 Drag Crisis

Before we close this chapter, we take a closer look at how the mean drag coefficient varies with the Reynolds number (Fig. 2.11). At very small Reynolds numbers, the mean drag coefficient has a large value (more than 10). As the Reynolds number increases from very small values, the mean drag coefficient decreases until $Re \sim 300$, where it reaches a value of close to $\overline{C_D} = 1.2$ and then it stays more or less constant for a wide range of Reynolds numbers up to $Re = 300{,}000$. At this Reynolds number, there is a sudden drop of the drag coefficient to much smaller values, close to $\overline{C_D} = 0.25$. This might seem surprising, as the mean drag coefficient did not change much for a wide range of Reynolds numbers before $Re = 300{,}000$. We call this sudden drop in the mean drag coefficient, the *drag crisis*. To understand what causes the drag crisis, we should remember our discussion on the wake of the cylinder versus the Reynolds number in Sect. 3 and remind ourselves of what happened at $Re = 300{,}000$: The separation point moves from $\theta = 80°$ to $\theta = 140°$, resulting in a much narrower wake for $Re > 300{,}000$. A smaller wake means a smaller range with an asymmetric pressure distribution with respect to a

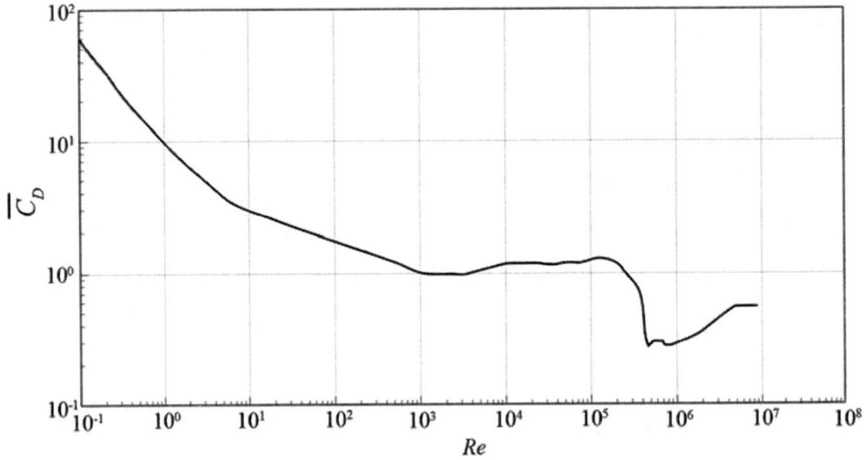

Fig. 2.11 Mean drag coefficient versus Reynolds number. (Figure reproduced from Sumer & Fredsoe, 2006)

vertical line that goes through the center of the cylinder, which means a smaller mean drag force. After the drag crisis, the drag coefficient stays around $\overline{C_D} = 0.2$ and then increases gradually for $Re > 1{,}500{,}000$ to a value close to $\overline{C_D} = 0.5$ and stays there for $Re > 4{,}500{,}000$.

Chapter 3
Vibrations of One and Two Degree-of-Freedom Systems

1 Introduction

In this chapter, we focus on the motion of the structure and study how it responds to an external force. In particular, we study the classical vibrations problem in which a mass is supported by a spring and a damper (known as a mass-spring-damper system) and is free to oscillate. Initially, we consider the case where there is only one mass (Fig. 3.1) and it can oscillate only in one direction (the y-direction in the figure) and therefore the system is called a one-degree-of-freedom (1 DOF) system. Later, in this chapter we consider a 2 DOF system. There is no fluid involved in a classical vibrations problem. The classical vibrations problem is discussed in several vibrations book (e.g., Thomson and Dahleh (1998), Rao (2011) and Inman (2014)) Here, we will focus on the parts that we will need in the future chapters of this book. In the very last section of this chapter, we will discuss the response of a mass-spring-damper system when it is placed in otherwise still fluid to study how the presence of fluid around a vibrating system affects its motion. We call this "otherwise still fluid" to emphasize that before the structure starts to oscillate, the fluid is at rest and it starts moving only because the structure is moving. This will be the first FSI system that we will study in this book.

2 An Undamped System

As a first step, we assume that the damping in the mass-spring-damper system of Fig. 3.1 is negligible. This is then an undamped 1 DOF system, whose equation of motion is:

Y. Modarres-Sadeghi, *Introduction to Fluid-Structure Interactions*,
https://doi.org/10.1007/978-3-030-85884-1_3

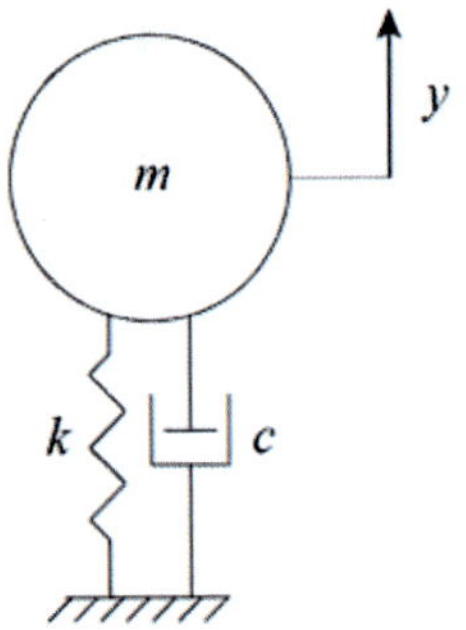

Fig. 3.1 A 1 DOF mass-spring-damper system. Vibrations are allowed only in the y-direction and are restricted in any other direction

$$m\ddot{y} + ky = 0, \tag{3.1}$$

which is the simplest equation that can describe a vibratory system. Equation (3.1) can be rewritten as

$$\ddot{y} + \frac{k}{m}y = 0, \tag{3.2}$$

where dot stands for the derivative with respect to time.

Eq. (3.2) is a second-order ordinary differential equation (ODE), and we need to know its appropriate initial conditions to be able to solve it. To decide on the initial conditions, we can think of what we need to do to a mass-spring system for it to start oscillating. We can either give an initial displacement to the mass, or an initial velocity to make it move. Or we can do both: give an initial displacement and an initial velocity. This means that the general initial conditions for this system are an initial displacement:

$$y(0) = y_0, \tag{3.3}$$

and an initial velocity:

$$\dot{y}(0) = v_0. \tag{3.4}$$

To solve Eq. (3.2) with the initial conditions of (3.3) and (3.4), we assume that the solution has a general form of

$$y(t) = Y \sin(\omega t + \phi). \tag{3.5}$$

In this general solution, there are three unknowns: Y, ω, and ϕ. We plug this general form of solution into the equation of motion (Eq. (3.1)) to find:

$$-\omega^2 Y \sin(\omega t + \phi) + \frac{k}{m} Y \sin(\omega t + \phi) = 0. \tag{3.6}$$

After dropping $Y \sin(\omega t + \phi)$, which is in common in both terms, we find:

$$-\omega^2 + \frac{k}{m} = 0, \tag{3.7}$$

and therefore

$$\omega = \sqrt{\frac{k}{m}} \equiv \omega_n. \tag{3.8}$$

We use symbol ω_n for this constant and we call it the *natural frequency* of the system. We call this the "natural" frequency, because it comes with the system naturally: it is a function of the system parameters only, and it is independent from any external force that could act on the system. Once we know the system, we know its mass and its stiffness, and this is all we need to know its natural frequency. It is called a natural "frequency" because it has the units of frequency (radians/second) and it shows the frequency at which the system would want to oscillate forever, given any initial condition. We should remember that the unit for ω_n is rad/s, not Hz, and we can relate it to the natural frequency in Hz (f_n) as $\omega_n = 2\pi f_n$.

To relate the other two unknowns, Y and ϕ, to the initial conditions and ω_n, we plug the initial conditions into the general solution of the equation of motion (Eq. (3.5)). By doing this, we get:

$$y(0) = y_0 = Y \sin \phi, \tag{3.9}$$

and

$$\dot{y}(0) = v_0 = Y\omega_n \cos \phi. \tag{3.10}$$

We divide (3.9) by (3.10):

$$\frac{y_0}{v_0} = \frac{Y \sin \phi}{Y\omega_n \cos \phi} = \frac{1}{\omega_n} \tan \phi \Rightarrow \tan \phi = \frac{y_0}{v_0}\omega_n, \tag{3.11}$$

and find that

$$\phi = \tan^{-1}\left(\frac{y_0}{v_0}\omega_n\right). \tag{3.12}$$

Then we find sinϕ from (3.9):

$$y_0 = Y \sin \phi \Rightarrow \sin \phi = \frac{y_0}{Y}, \tag{3.13}$$

and cosϕ from (3.10):

$$v_0 = Y\omega_n \cos\phi \Rightarrow \cos\phi = \frac{v_0}{Y\omega_n}, \quad (3.14)$$

and then we use the fact that $\sin^2\phi + \cos^2\phi = 1$ to find:

$$\left(\frac{y_0}{Y}\right)^2 + \left(\frac{v_0}{Y\omega_n}\right)^2 = 1, \quad (3.15)$$

which will give us:

$$Y = \frac{\sqrt{y_0^2\omega_n^2 + v_0^2}}{\omega_n}. \quad (3.16)$$

Then, the general solution (Eq. (3.5)) can be written as

$$y(t) = \frac{\sqrt{y_0^2\omega_n^2 + v_0^2}}{\omega_n} \sin\left(\omega_n t + \tan^{-1}\left(\frac{y_0}{v_0}\omega_n\right)\right). \quad (3.17)$$

This is a sinusoidal response whose amplitude depends on the initial conditions and whose frequency is equal to the natural frequency of the system. This means that if we change the initial conditions, then we will get another sinusoidal response with the same frequency, but with a different amplitude. For a given set of initial conditions, the amplitude of oscillations stays constant throughout the oscillations (Fig. 3.2), which means that the system will never stop oscillating, since there is no

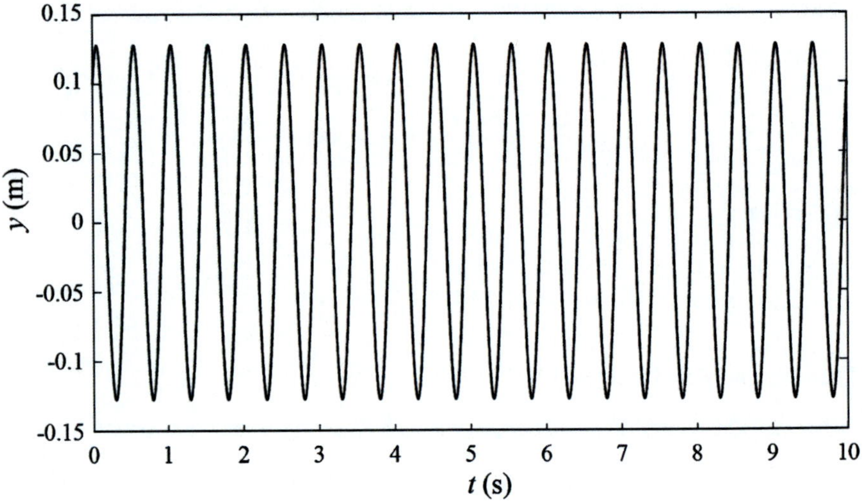

Fig. 3.2 Oscillations of an undamped 1 DOF system with $f_n = 2$ Hz and $y_0 = 0.1$ m, $v_0 = 1$ m/s

source for dissipating the energy. In the following section, we will consider a more realistic case, where a structural damping exists in the system and damps the oscillations.

3 A Damped System

When a structural damping does exist, the equation of motion takes the following form:

$$m\ddot{y} + c\dot{y} + ky = 0, \tag{3.18}$$

where c is a damping coefficient. The term $c\dot{y}$ represents a damping force proportional to the velocity of the structure. This is called a *viscous damping*. A viscous damping represents the internal dissipation that we normally observe in structures and its force is linearly proportional to the velocity of the structure. We should remember that a viscous damping does not represent the friction that is exerted on a body that is in contact with a rough surface. A coulomb damping, which is a nonlinear damping, represents the friction force. Here, we do not discuss the coulomb damping and only consider the viscous damping.

All the terms in Eq. (3.18) have units of force, N. It is easy to remember the units for mass, $[m] = kg$. If we cannot remember the units of the damping coefficient and stiffness, we can use the equation of motion and derive them. The unit for the damping coefficient is

$$[c] = \frac{N}{\frac{m}{s}} = \frac{kg\frac{m}{s^2}}{\frac{m}{s}} = \frac{kg}{s}, \tag{3.19}$$

and the unit for stiffness is

$$[k] = \frac{N}{m} = \frac{kg\frac{m}{s^2}}{m} = \frac{kg}{s^2}. \tag{3.20}$$

Before solving Eq. (3.18), we will take some steps to rewrite it in a more useful format. First, we divide Eq. (3.18) by m

$$\ddot{y} + \frac{c}{m}\dot{y} + \frac{k}{m}y = 0. \tag{3.21}$$

We know that

$$\frac{k}{m} \equiv \omega_n^2, \tag{3.22}$$

and we can rewrite the damping term, c/m, as

$$\frac{c}{m} = \frac{2}{2}\frac{c}{\sqrt{m}\sqrt{m}} \times \frac{\sqrt{k}}{\sqrt{k}} = 2\frac{c}{2\sqrt{km}}\sqrt{\frac{k}{m}}. \tag{3.23}$$

We define $\zeta \equiv c/(2\sqrt{km})$ and write the damping term as

$$\frac{c}{m} = 2\zeta\omega_n, \tag{3.24}$$

where ζ is a dimensionless parameter that represents the structural damping, and we call it the *damping ratio*. We can show that ζ is dimensionless:

$$[\zeta] = \left[\frac{c}{2\sqrt{km}}\right] = \frac{\frac{kg}{s}}{2\sqrt{\frac{kg}{s^2}kg}} = \frac{\frac{kg}{s}}{2\sqrt{\left(\frac{kg}{s}\right)^2}} = [1], \tag{3.25}$$

and we can rewrite it as

$$\zeta = \frac{c}{2\sqrt{km}} = \frac{c}{2m\omega_n} \equiv \frac{c}{c_{cr}}, \tag{3.26}$$

where $c_{cr} = 2m\omega_n$ is called the *critical damping*.

Now we can write the equation of motion (Eq. (3.18)) in terms of ω_n and ζ:

$$\ddot{y} + 2\zeta\omega_n\dot{y} + \omega_n^2 y = 0. \tag{3.27}$$

This form is more useful than the original form of Eq. (3.18), because in this form we can discuss the system response in terms of its natural frequency, ω_n, and damping ratio, ζ, which makes the discussions more universal.

For Eq. (3.27), we assume the initial conditions of

$$y(0) = y_0, \tag{3.28}$$

and

$$\dot{y}(0) = v_0. \tag{3.29}$$

To solve the equation, we assume that the general solution is

$$y(t) = Ye^{\lambda t}, \tag{3.30}$$

in which Y and λ are unknowns. We plug Eq. (3.30) into the equation of motion (Eq. (3.27)) to get:

$$\lambda^2 Ye^{\lambda t} + 2\zeta\omega_n \lambda Ye^{\lambda t} + \omega_n^2 Ye^{\lambda t} = 0, \tag{3.31}$$

and then drop $Ye^{\lambda t}$ to find a second-order algebraic equation:

$$\lambda^2 + 2\zeta\omega_n \lambda + \omega_n^2 = 0, \tag{3.32}$$

which we can solve to find:

$$\lambda = -\zeta\omega_n \pm \sqrt{\zeta^2\omega_n^2 - \omega_n^2}, \tag{3.33}$$

or

$$\lambda = -\zeta\omega_n \pm \omega_n\sqrt{\zeta^2 - 1}. \tag{3.34}$$

Three possibilities are normally considered, depending on the value of the damping ratio: (i) when $\zeta < 1$ (an underdamped case); (ii) when $\zeta = 1$ (a critically damped case), and (iii) when $\zeta > 1$ (an overdamped case). Here we consider the underdamped case only, because this is the only case that we will use later in this book, and this is the only case that will give us oscillatory motion. For an underdamped case, $\zeta < 1$, we can write λ as

$$\lambda = -\zeta\omega_n \pm \omega_n\sqrt{(-1)(1 - \zeta^2)}, \tag{3.35}$$

or

$$\lambda = -\zeta\omega_n \pm i\omega_n\sqrt{1 - \zeta^2}, \tag{3.36}$$

where $i = \sqrt{-1}$.

Then we plug λ into the general solution (Eq. (3.30)) and we find:

$$y(t) = a_1 e^{\left(-\zeta\omega_n + i\omega_n\sqrt{1-\zeta^2}\right)t} + a_2 e^{\left(-\zeta\omega_n - i\omega_n\sqrt{1-\zeta^2}\right)t}, \tag{3.37}$$

or

$$y(t) = e^{-\zeta\omega_n t}\left(a_1 e^{+i\omega_n\sqrt{1-\zeta^2}\, t} + a_2 e^{-i\omega_n\sqrt{1-\zeta^2}\, t}\right). \tag{3.38}$$

The term $\omega_n\sqrt{1 - \zeta^2}$ shows up in both terms. This is a term that depends on the system's natural frequency, ω_n, and damping ratio, ζ. We call this the *damped natural frequency* and we use ω_d to refer to it. Then:

$$\omega_n\sqrt{1-\zeta^2} \equiv \omega_d, \tag{3.39}$$

which will give the following form of the general solution:

$$y(t) = e^{-\zeta\omega_n t}(a_1 e^{+i\omega_d t} + a_2 e^{-i\omega_d t}). \tag{3.40}$$

We know that the Euler's formula relates the exponentials with an imaginary exponent to the sine and cosine functions, $e^{i\phi} = \cos\phi + i\sin\phi$. If we use the Euler's formula for the two exponentials with imaginary exponents in the general solution, then:

$$y(t) = e^{-\zeta\omega_n t}(a_1(\cos\omega_d t + i\sin\omega_d t) + a_2(\cos\omega_d t - i\sin\omega_d t)), \tag{3.41}$$

or

$$y(t) = e^{-\zeta\omega_n t}((a_1 + a_2)\cos\omega_d t + i(a_1 - a_2)\sin\omega_d t). \tag{3.42}$$

We assume:

$$\begin{aligned} a_1 + a_2 &= A_1, \\ i(a_1 - a_2) &= A_2, \end{aligned} \tag{3.43}$$

which gives us

$$y(t) = e^{-\zeta\omega_n t}(A_1\cos\omega_d t + A_2\sin\omega_d t), \tag{3.44}$$

and then we assume

$$\begin{aligned} A_1 &= Y\sin\phi, \\ A_2 &= Y\cos\phi, \end{aligned} \tag{3.45}$$

which results in

$$y(t) = e^{-\zeta\omega_n t}(Y\sin\phi\cos\omega_d t + Y\cos\phi\sin\omega_d t), \tag{3.46}$$

which can be rewritten as:

$$y(t) = Ye^{-\zeta\omega_n t}\sin(\omega_d t + \phi). \tag{3.47}$$

Eq. (3.47) is the solution for Eq. (3.27). In this solution, Y and ϕ are unknowns, and they can be found using the initial conditions. If we apply the initial conditions to the solution, we will find:

$$y_0 = Y \sin \phi, \tag{3.48}$$

for the initial displacement, and

$$v_0 = -Y\zeta\omega_n \sin \phi + Y\omega_d \cos \phi, \tag{3.49}$$

for the initial velocity, which we rewrite as

$$v_0 + \zeta\omega_n y_0 = Y\omega_d \cos \phi. \tag{3.50}$$

Then we divide (3.48) by (3.50) to get:

$$\frac{y_0}{v_0 + \zeta\omega_n y_0} = \frac{\tan \phi}{\omega_d}, \tag{3.51}$$

and

$$\phi = \tan^{-1} \frac{\omega_d y_0}{v_0 + \zeta\omega_n y_0}. \tag{3.52}$$

Then we find sinϕ from (3.48) and cosϕ from (3.50) and use $\sin^2\phi + \cos^2\phi = 1$ to get:

$$\left(\frac{y_0}{Y}\right)^2 + \left(\frac{v_0 + \zeta\omega_n y_0}{Y\omega_d}\right)^2 = 1, \tag{3.53}$$

and therefore

$$Y = \sqrt{{y_0}^2 + \left(\frac{v_0 + \zeta\omega_n y_0}{\omega_d}\right)^2}. \tag{3.54}$$

Equations (3.52) and (3.54) give the amplitude and phase of the general solution in terms of the system parameters and the initial conditions, leading to the solution for a 1 DOF damped system:

$$y(t) = \sqrt{{y_0}^2 + \left(\frac{v_0 + \zeta\omega_n y_0}{\omega_d}\right)^2} e^{-\zeta\omega_n t} \sin\left(\omega_d t + \tan^{-1} \frac{\omega_d y_0}{v_0 + \zeta\omega_n y_0}\right). \tag{3.55}$$

Based on this solution, the response is a sinusoidal one that starts at a nonzero value and decays exponentially to zero. A sample of this type of response is shown in Fig. 3.3. How fast the solution goes to zero depends on $\zeta\omega_n$: a larger $\zeta\omega_n$ implies a faster decay and a smaller one a slower decay.

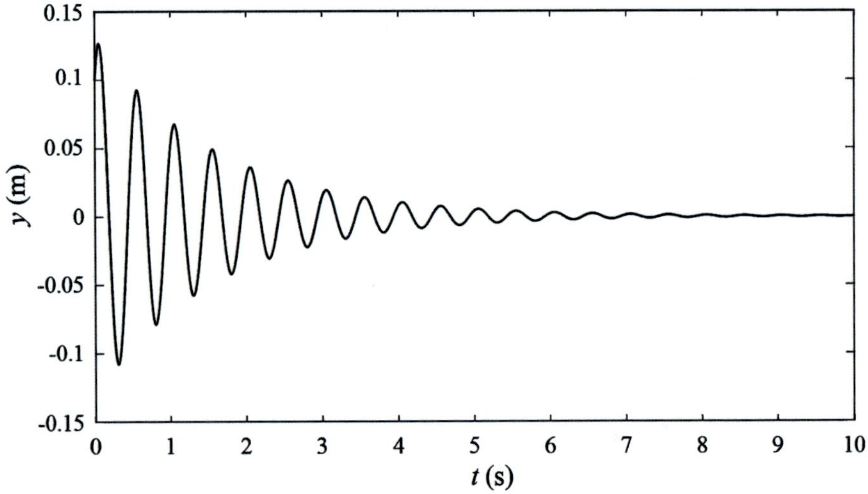

Fig. 3.3 Oscillations of a damped 1 DOF system with $f_n = 2$ Hz, $\zeta = 0.05$, and $y_0 = 0.1$ m, $v_0 = 1$ m/s

If $\zeta = 0$, then Eq. (3.55) becomes the same as the solution we found for the undamped system (Eq. (3.17)).

For the critically damped and overdamped cases, there will not be any imaginary exponent in the exponential in Eq. (3.30) and therefore no sinusoidal response will be observed. (Remember the Euler's formula).

4 Measuring the Structural Damping

Before we move on, we should discuss a practical matter and that is the way we can measure the structural damping in a system. Measuring mass and stiffness (spring constant) is rather easy: We weigh the system and find its mass and we attach a known mass to the spring and measure the distance it stretches and calculate its stiffness. Measuring the damping coefficient is not as trivial. To measure the damping coefficient, if we conduct an experiment on a mass-spring-damper system and we measure the system's displacement, then we find a plot similar to that of Fig. 3.3. This plot must follow the general solution (Eq. (3.47)).

To measure ζ, we need to fit a curve to the magnitude of the decaying response of Fig. 3.3. Since the decay is caused by the exponential part of the response, $e^{-\zeta\omega_n t}$, the curve that we will fit should have the same form, and it will give us the value of $\zeta\omega_n$. We can also plot the logarithm of the response, $\ln(y)$, to avoid dealing with the exponential and then find the slope of the resulting line to find $\zeta\omega_n$. Then if we use the same time history of Fig. 3.3 and find its frequency content by finding the Fourier transform of the signal, we will find a single peak in the frequency plot. This peak

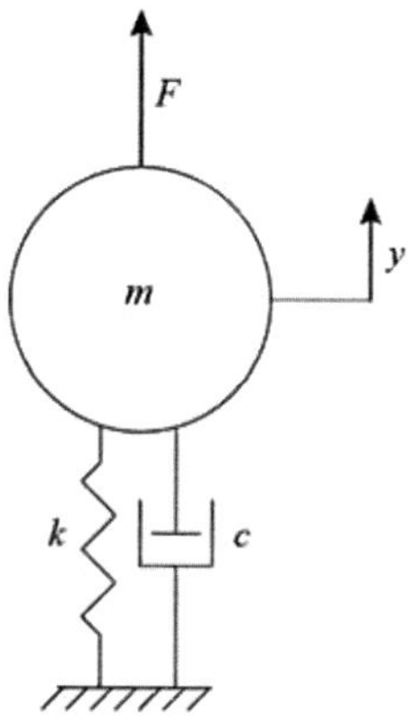

Fig. 3.4 A mass-spring-damper with an external force

corresponds to the damped natural frequency of the system, f_d, since the damping does exist in our system, and it influences our experimentally-measured frequency. Then from the measurement we have found $\omega_d = 2\pi f_d$, and $\zeta\omega_n$. The goal is to find ζ. We can write $\zeta\omega_n$ as $\zeta\omega_n = \zeta\omega_d/\sqrt{1-\zeta^2}$. Since we have measured $\zeta\omega_n$, it is a known constant: $\zeta\omega_n \equiv Z$, and therefore

$$Z = \zeta\frac{\omega_d}{\sqrt{1-\zeta^2}}, \tag{3.56}$$

and then we solve for ζ:

$$\zeta = \frac{Z}{\sqrt{Z + \omega_d^2}}. \tag{3.57}$$

The damping ratio that we find here must be smaller than 1, because it is for an undamped system. In cases of very small damping, e.g., $\zeta \simeq 0.01$, f_n and f_d are practically the same.

5 Harmonic Excitation of Damped Systems

Here, we consider the case where a mass-spring-damper experiences an external force. This takes us one step closer to our main topic, which is the case where a structure is forced externally by fluid forces. For now, we assume that the external force is independent from the structure's motion. Later on, when the source of the external force is flow, the magnitude of the external force will depend on the system's response.

The equation of motion for a 1 DOF system experiencing an external force is

$$m\ddot{y} + c\dot{y} + ky = F(t). \tag{3.58}$$

To reiterate, we assume that the external force is independent from the system's oscillations. This force, $F(t)$, could be any function of time. Here, we assume a sinusoidal function with an arbitrary frequency, ω:

$$m\ddot{y} + c\dot{y} + ky = F_0 \cos \omega t. \tag{3.59}$$

We can rewrite Eq. (3.59) as:

$$\ddot{y} + 2\zeta\omega_n\dot{y} + \omega_n^2 y = \frac{F_0}{m} \cos \omega t \equiv f_0 \cos \omega t. \tag{3.60}$$

This is a second-order linear ODE with a nonzero right-hand side. In the previous cases that we discussed, there was no external force and the right-hand side of the equation was zero. In the present case, the external force appears on the right-hand side and makes it nonzero. The solution for this system, therefore, has two parts: homogenous, y_h, and particular, y_p:

$$y = y_h + y_p. \tag{3.61}$$

The homogenous solution is the solution of the system without any external force, i.e., with a zero right-hand side. That is the "natural" response of the damped system, which will go to zero after some time, as we discussed in the previous section, leaving the system with the particular solution only. The particular solution, as the name suggests, is particular to the external force that acts on the system. Physically, that is the solution that remains as long as the external force is acting on the system. For a damped system, the homogeneous solution damps out by the end of the "transient" stage and the particular solution defines the "steady-state" response.

The homogenous solution is the solution to

$$\ddot{y} + 2\zeta\omega_n\dot{y} + \omega_n^2 y = 0, \tag{3.62}$$

which is

$$y_h = Ye^{-\zeta\omega_n t} \sin(\omega_d t + \phi), \tag{3.63}$$

as we discussed before.

The particular solution must satisfy Eq. (3.60) when the right-hand side is not zero. Obviously, this solution will change with different forcing functions. For the case we have considered here, the external force has the form of $f_0 \cos \omega t$, which implies that the particular solution should have a similar form to be able to satisfy the equation. We pick the following form for the particular solution:

$$y_p = Y_P \cos(\omega t - \theta). \tag{3.64}$$

Note that the frequency in the particular solution is ω and not ω_d. This makes the particular solution have a frequency equal to the excitation frequency, which is in line with the role of the particular solution: it exists only because there is an external force and therefore its frequency should be equal to the frequency of the external force. In this form of solution, Y_P represents the amplitude of the steady state response and the angle θ represents the phase difference between the response and the external force. This phase difference exists due to the damping term. Mathematically, if we have a cosine on the right-hand side and we would like to satisfy the left-hand side with a damping term, then we must create a term with a sine on the left-hand side with our solution. This is what the phase difference does in this case: it gives us the possibility of creating a sine term. We still do not know the values of Y_P and θ, but we will find them soon. It will not be surprising if we find that they depend on the system parameters (i.e., ω_n and ζ), as well as the external force amplitude and frequency (f_0 and ω).

Let's plug the particular solution (Eq. (3.64)) into the equation of motion (Eq. (3.60)):

$$\begin{aligned}&-\omega^2 Y_P \cos(\omega t - \theta) - 2\zeta\omega\omega_n Y_P \sin(\omega t - \theta) + \omega_n^2 Y_P \cos(\omega t - \theta)\\ &\quad = f_0 \cos \omega t.\end{aligned} \tag{3.65}$$

Collecting $\sin(\omega t - \theta)$ and $\cos(\omega t - \theta)$ terms on the left-hand side,

$$\left(\omega_n^2 - \omega^2\right) Y_P \cos(\omega t - \theta) - 2\zeta\omega\omega_n Y_P \sin(\omega t - \theta) = f_0 \cos \omega t, \tag{3.66}$$

and then expanding $\sin(\omega t - \theta)$ and $\cos(\omega t - \theta)$ terms,

$$\begin{aligned}&\left(\omega_n^2 - \omega^2\right) Y_P(\cos \omega t \cos \theta + \sin \omega t \sin \theta)\\ &\quad - 2\zeta\omega\omega_n Y_P(\sin \omega t \cos \theta - \cos \omega t \sin \theta)\\ &\quad = f_0 \cos \omega t,\end{aligned} \tag{3.67}$$

and again collecting $\sin\omega t$ and $\cos\omega t$ terms on the left-hand side, we find:

$$\begin{aligned}&\left[\left(\omega_n^2 - \omega^2\right) Y_P \cos \theta + 2\zeta\omega\omega_n Y_P \sin \theta\right] \cos \omega t\\ &+\left[\left(\omega_n^2 - \omega^2\right) Y_P \sin \theta - 2\zeta\omega\omega_n Y_P \cos \theta\right] \sin \omega t\\ &= f_0 \cos \omega t,\end{aligned} \tag{3.68}$$

from where we find two equations: one from the coefficient of $\cos\omega t$ and the other from the coefficient of $\sin\omega t$, resulting in the following set of equations:

$$\begin{cases} \left(\omega_n^2 - \omega^2\right) Y_P \cos\theta + 2\zeta\omega\omega_n Y_P \sin\theta = f_0, & \text{(a)} \\ \left(\omega_n^2 - \omega^2\right) Y_P \sin\theta - 2\zeta\omega\omega_n Y_P \cos\theta = 0. & \text{(b)} \end{cases} \tag{3.69}$$

We need to use these equations and find Y_P and θ in terms of the system parameters and the external force's amplitude and frequency. From Eq. (3.69b), we find:

$$\frac{\sin\theta}{\cos\theta} = \frac{2\zeta\omega\omega_n}{\omega_n^2 - \omega^2} \equiv \tan\theta, \tag{3.70}$$

and then:

$$\theta = \tan^{-1}\frac{2\zeta\omega\omega_n}{\omega_n^2 - \omega^2}. \tag{3.71}$$

This is the phase difference in terms of the known parameters. Note that the phase difference does not depend on the amplitude of the external force. Then, multiplying Eq. (3.69a) by cosθ and Eq. (3.69b) by sinθ and adding them to each other, we find:

$$\left(\omega_n^2 - \omega^2\right) Y_P = f_0 \cos\theta, \tag{3.72}$$

and

$$Y_P = \frac{f_0 \cos\theta}{\omega_n^2 - \omega^2}. \tag{3.73}$$

We write cosθ in terms of tanθ, since we have already found tanθ in terms of the system parameters in Eq. (3.70). To relate them, we remember from the trigonometric relations that:

$$\cos^2\theta = \frac{1}{1 + \tan^2\theta}, \tag{3.74}$$

and then using Eq. (3.70), we find:

$$\cos^2\theta = \frac{1}{1 + \left(\frac{2\zeta\omega\omega_n}{\omega_n^2 - \omega^2}\right)^2} = \frac{\left(\omega_n^2 - \omega^2\right)^2}{\left(\omega_n^2 - \omega^2\right)^2 + \left(2\zeta\omega\omega_n\right)^2}, \tag{3.75}$$

or

$$\cos\theta = \frac{\left(\omega_n^2 - \omega^2\right)}{\sqrt{\left(\omega_n^2 - \omega^2\right)^2 + \left(2\zeta\omega\omega_n\right)^2}}, \tag{3.76}$$

which makes Eq. (3.73) become:

$$Y_P = \frac{f_0}{\sqrt{\left(\omega_n^2 - \omega^2\right)^2 + \left(2\zeta\omega\omega_n\right)^2}}. \tag{3.77}$$

This is the magnitude of the particular solution in terms of the known parameters. Note that it depends both on the magnitude and frequency of the external force and the natural frequency and damping ratio of the system. Eqs. (3.71) and (3.77) give us what we need to know the particular solution of the system. In both of these relations, ω and ω_n show up. We can rewrite these relations such that the ratio of the external frequency and the natural frequency, ω / ω_n, shows up in them. Then for Eq. (3.71) we will have:

$$\theta = \tan^{-1} \frac{2\zeta \frac{\omega}{\omega_n}}{1 - \left(\frac{\omega}{\omega_n}\right)^2} \equiv \tan^{-1} \frac{2\zeta r}{1 - r^2}, \tag{3.78}$$

where we call $r = \omega/\omega_n$ the frequency ratio, and for Eq. (3.77) we have

$$Y_P = \frac{f_0}{\omega_n^2 \sqrt{\left(1 - \left(\frac{\omega}{\omega_n}\right)^2\right)^2 + \left(2\zeta \frac{\omega}{\omega_n}\right)^2}}, \tag{3.79}$$

which we can rewrite as:

$$\frac{Y_P\omega_n^2}{f_0} = \frac{1}{\sqrt{(1 - r^2)^2 + (2\zeta r)^2}}. \tag{3.80}$$

Using Eq. (3.60), the left-hand side of Eq. (3.80) can be written as

$$\frac{Y_P\omega_n^2}{f_0} = \frac{Y_P\omega_n^2 m}{F_0} = \frac{Y_P \frac{k}{m} m}{F_0} = \frac{Y_P k}{F_0}, \tag{3.81}$$

where F_0/k is the displacement that we observe when a constant force F_0 acts on the system. We call this a static deflection, Y_{St}, which makes Eq. (3.80) read as:

$$\frac{Y_P}{Y_{St}} = \frac{1}{\sqrt{(1 - r^2)^2 + (2\zeta r)^2}}. \tag{3.82}$$

Eqs. (3.78) and (3.82) give us θ and Y_P in terms of the system parameters and the external forcing. On the right-hand side of both of these relations, we only see two dimensionless parameters: r and ζ. This gives us the ability to plot these for a range of r and ζ values and see how any 1 DOF vibratory system behaves under external sinusoidal forcing. These plots are shown in Fig. 3.5, for both the amplitude and the phase. For $r = 0$, we have a static external force, which results in a static deflection equal to Y_{St}, leading to $\frac{Y_P}{Y_{St}} = 1$. As we increase the external frequency, i.e., increasing r, the dimensionless amplitude deviates from 1. For $\zeta = 0$, the amplitude increases to infinity as r is increased from 0 to 1 and then decreases for $r > 1$. When $r = 1$, the external frequency and the natural frequency are equal. This is called *resonance*. The amplitude of oscillations at resonance is equal to (plug $r = 1$ into Eq. (3.82)):

$$Y_{P\_res} = \frac{F_0}{k}\frac{1}{2\zeta}, \tag{3.83}$$

which goes to infinity if there is no structural damping ($\zeta = 0$), but has a finite value if there is any structural damping. Since there is practically no case with $\zeta = 0$ in real life, we do not see oscillations whose amplitude goes to infinity. It is easy to imagine, and it also follows from Eq. (3.83), that a larger structural damping results in a smaller amplitude of resonance. This is also clear in Fig. 3.5a.

If there is no structural damping, θ jumps from 0° to 180° at resonance. With a nonzero structural damping, however, the transition from 0° to 180° occurs more gradually (Fig. 3.5b).

It is important to note that the maximum amplitude of oscillations does not occur at resonance. This is also clear in Fig. 3.5a: for any ζ, the peak is not at $r = 1$—it is close to it but to its left. To find the maximum amplitude, we must find the value of r for which the derivative of the amplitude with respect to r is equal to zero From Eq. (3.82):

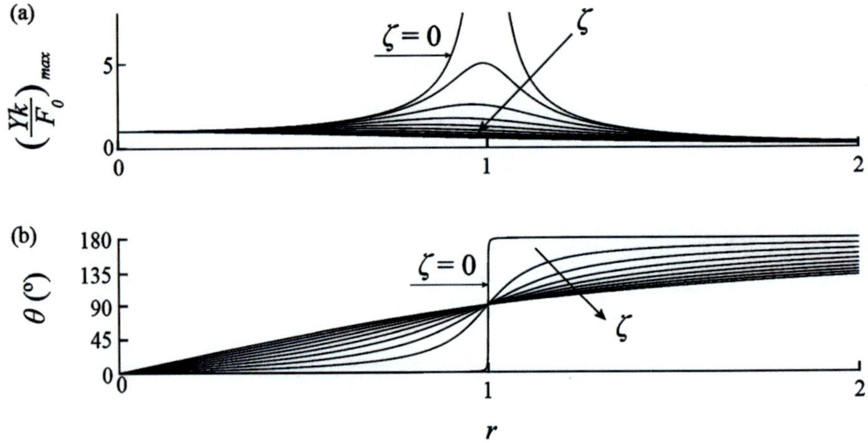

Fig. 3.5 (**a**) The dimensionless amplitude and (**b**) the phase angle versus frequency ratio for a 1 DOF vibratory system

$$\frac{dY_P}{dr} = 0 \Rightarrow -4r(1 - r^2) + 8r\zeta^2 = 0, \tag{3.84}$$

resulting in:

$$r = \sqrt{1 - 2\zeta^2}, \tag{3.85}$$

which is clearly smaller than $r = 1$ (the value at resonance). Here, r is physically meaningful only when it is a positive real number. Therefore, we will have a local maximum only if

$$1 - 2\zeta^2 \geq 0 \quad \Rightarrow \quad \zeta \leq \frac{1}{\sqrt{2}}, \tag{3.86}$$

which means that a maximum amplitude of oscillations exists only for $\zeta \leq 1/\sqrt{2}$, and no maximum exists for larger ζ values. This is clear in Fig. 3.5a that for $\zeta > 1/\sqrt{2}$, the amplitude of oscillations decreases with increasing r, and as a result, there is no local maximum for $r > 0$. In summary, for the maximum amplitude we have:

$$\begin{array}{ll} \left(\frac{Y_P k}{F_0}\right)_{\text{Max}} = \frac{1}{2\zeta\sqrt{1-\zeta^2}} & \zeta \leq \frac{1}{\sqrt{2}} \\ \text{No Maximum} & \zeta > \frac{1}{\sqrt{2}} \end{array} \tag{3.87}$$

Overall, Fig. 3.5 summarizes the response of any 1 DOF system, with a given natural frequency and damping ratio, to an external sinusoidal force, when the external force has a constant amplitude and a constant frequency.

6 Systems with Two Degrees of Freedom

In this section, we consider a system with two degrees of freedom. The degree of freedom is the set of independent coordinates that specifies the position of a system completely. Then when we need two independent coordinates to identify the system's response, we say that we have a 2 DOF system. This means that we need two equations to describe a 2 DOF system. A sample 2 DOF system is shown in Fig. 3.6. This system has two masses, m_1 and m_2, three springs, k_1, k_2, and k_3, and no damping, only because we assume there is no damping. We can hold the first mass in place and make sure that it does not move, but even then, the second mass can still oscillate. If instead, we hold the second mass in place, then the first mass can still oscillate. This shows that the two displacements are independent from each other, and that we have a 2 DOF system. In order to know the system's response, we need

Fig. 3.6 A 2 DOF vibratory system

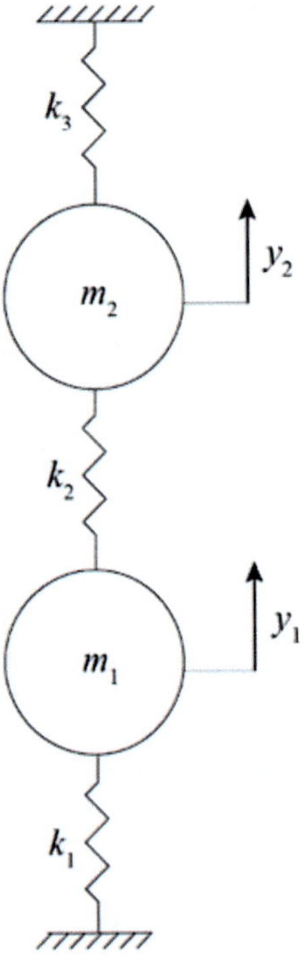

to know how y_1 and y_2 change with time. We have to keep in mind that to have a 2 DOF system we do not need to have 2 masses. As an example, we can have one mass, which is free to oscillate in a plane. That also gives us a 2 DOF system.

For the system of Fig. 3.6, the equations of motion are:

$$\begin{cases} m_1\ddot{y}_1 + (k_1 + k_2)y_1 - k_2y_2 = 0, \\ m_2\ddot{y}_2 - k_2y_1 + (k_2 + k_3)y_2 = 0, \end{cases} \tag{3.88}$$

which are two coupled ODEs. They are coupled because y_1 and y_2 exist in both of them, and we cannot solve one of these equations without solving the other one. We can write this set of equations in a matrix form as:

$$\begin{bmatrix} m_1 & 0 \\ 0 & m_2 \end{bmatrix} \begin{Bmatrix} \ddot{y}_1 \\ \ddot{y}_2 \end{Bmatrix} + \begin{bmatrix} k_1 + k_2 & -k_2 \\ -k_2 & k_2 + k_3 \end{bmatrix} \begin{Bmatrix} y_1 \\ y_2 \end{Bmatrix} = \begin{Bmatrix} 0 \\ 0 \end{Bmatrix}, \tag{3.89}$$

or

$$\mathbf{M}\ddot{\mathbf{y}} + \mathbf{K}\mathbf{y} = \mathbf{0}, \tag{3.90}$$

in which **M** and **K** are called the mass matrix and stiffness matrix, respectively. The form of this matrix equation is the same as the form of the equation of motion for a 1 DOF vibratory system (Eq. (3.1)): we have an inertia term, $\mathbf{M}\ddot{\mathbf{y}}$, and a stiffness term, $\mathbf{K}\mathbf{y}$. The method of solution is very similar too. We assume a general form of solution for the system as:

$$\mathbf{y} = \mathbf{Y}e^{i\omega t}, \tag{3.91}$$

where $\mathbf{y} = \begin{Bmatrix} y_1 \\ y_2 \end{Bmatrix}$ and $\mathbf{Y} = \begin{Bmatrix} Y_1 \\ Y_2 \end{Bmatrix}$. Then we plug this form back into the equations of motion (Eq. (3.89)) to get

$$\begin{bmatrix} m_1 & 0 \\ 0 & m_2 \end{bmatrix}\begin{Bmatrix} -Y_1\omega^2 e^{i\omega t} \\ -Y_2\omega^2 e^{i\omega t} \end{Bmatrix} + \begin{bmatrix} k_1 + k_2 & -k_2 \\ -k_2 & k_2 + k_3 \end{bmatrix}\begin{Bmatrix} Y_1 e^{i\omega t} \\ Y_2 e^{i\omega t} \end{Bmatrix} = \begin{Bmatrix} 0 \\ 0 \end{Bmatrix}, \tag{3.92}$$

and after dropping $e^{i\omega t}$, which is in common in all terms:

$$\begin{bmatrix} m_1 & 0 \\ 0 & m_2 \end{bmatrix}\begin{Bmatrix} -Y_1\omega^2 \\ -Y_2\omega^2 \end{Bmatrix} + \begin{bmatrix} k_1 + k_2 & -k_2 \\ -k_2 & k_2 + k_3 \end{bmatrix}\begin{Bmatrix} Y_1 \\ Y_2 \end{Bmatrix} = \begin{Bmatrix} 0 \\ 0 \end{Bmatrix}, \tag{3.93}$$

which gives us two equations:

$$\begin{Bmatrix} (-m_1\omega^2 + k_1 + k_2)Y_1 - k_2 Y_2 = 0, \\ -k_2 Y_1 + (-m_2\omega^2 + k_2 + k_3)Y_2 = 0, \end{Bmatrix} \tag{3.94}$$

or in matrix form:

$$\begin{bmatrix} -m_1\omega^2 + k_1 + k_2 & -k_2 \\ -k_2 & -m_2\omega^2 + k_2 + k_3 \end{bmatrix}\begin{Bmatrix} Y_1 \\ Y_2 \end{Bmatrix} = \begin{Bmatrix} 0 \\ 0 \end{Bmatrix}. \tag{3.95}$$

The trivial solution for this set of equations is $Y_1 = Y_2 = 0$, which means that the amplitudes of oscillations are zero, and therefore there are no oscillations. This is obviously not what we are looking for. In order to get the nontrivial solution of the equation, we have to make sure that the determinant of the matrix in Eq. (3.95) is equal to zero:

$$\begin{vmatrix} -m_1\omega^2 + k_1 + k_2 & -k_2 \\ -k_2 & -m_2\omega^2 + k_2 + k_3 \end{vmatrix} = 0, \tag{3.96}$$

which implies that the two equations are linearly dependent. Then Eq. (3.96) gives us:

$$\left(-m_1\omega^2 + k_1 + k_2\right)\left(-m_2\omega^2 + k_2 + k_3\right) - k_2^2 = 0, \tag{3.97}$$

or

$$m_1 m_2 \omega^4 - [(k_2 + k_3)m_1 + (k_1 + k_2)m_2]\omega^2 + k_1 k_2 + k_1 k_3 + k_2 k_3 = 0. \tag{3.98}$$

This is called the *characteristic equation*. It is a fourth-order algebraic equation in ω and a second-order in ω^2. If we solve it for ω^2, we will get two values: ω_1^2 and ω_2^2, which will give us four values for ω: $\pm\omega_1$ and $\pm\omega_2$. These frequencies are the natural frequencies of the system, and since we cannot have negative natural frequencies, then we pick the positive values of ω_1 and ω_2 as physically meaningful values for the natural frequencies. We use ω_1 to refer to the smaller value and ω_2 to refer to the larger value between the two. Then we call ω_1 the first natural frequency and ω_2 the second natural frequency.

Now that we know ω_1 and ω_2 we can use Eq. (3.95) to find the ratio between Y_1 and Y_2. Note that since initially we had three unknowns (ω, Y_1, and Y_2) and two equations in Eq. (3.95), all we can do is to find the ratio between Y_1 and Y_2 (Y_1/Y_2) corresponding to each of these natural frequencies. If we use the first line in Eq. (3.95), then we find:

$$\left(-m_1\omega_1^2 + k_1 + k_2\right)Y_1 - k_2 Y_2 = 0, \tag{3.99}$$

which gives us

$$\frac{Y_1}{Y_2} = \frac{k_2}{-m_1\omega_1^2 + k_1 + k_2}, \tag{3.100}$$

or if we use the second line in Eq. (3.95), then we find

$$-k_2 Y_1 + \left(-m_2\omega_1^2 + k_2 + k_3\right)Y_2 = 0, \tag{3.101}$$

which gives us:

$$\frac{Y_1}{Y_2} = \frac{-m_2\omega_1^2 + k_2 + k_3}{k_2}. \tag{3.102}$$

Both Eqs. (3.100) and (3.102) give us the ratio between the magnitudes of oscillations for mass 1 and mass 2, when the system oscillates at its first natural frequency, ω_1, and they both result in the same numerical value, i.e., if we get a ratio of 2 from Eq. (3.100), then we will get a ratio of 2 from Eq. (3.102) too. In the same way, we can plug ω_2 into the first or the second line of Eq. (3.95), and find the ratio between the two magnitudes when the system oscillates at its second natural frequency, ω_2. Then we can summarize these and say that we use r_1 to refer to the ratio of Y_1/Y_2 that corresponds to ω_1:

$$r_1 = \left(\frac{Y_1}{Y_2}\right)_1 = \frac{k_2}{-m_1\omega_1^2 + k_1 + k_2} = \frac{-m_2\omega_1^2 + k_2 + k_3}{k_2}, \tag{3.103}$$

and r_2 to refer to the ratio of $\frac{Y_1}{Y_2}$ that corresponds to ω_2:

$$r_2 = \left(\frac{Y_1}{Y_2}\right)_2 = \frac{k_2}{-m_1\omega_2^2 + k_1 + k_2} = \frac{-m_2\omega_2^2 + k_2 + k_3}{k_2}. \tag{3.104}$$

As an example, if in the system of Fig. 3.6, $m_1 = m_2 = m$ and $k_1 = k_2 = k_3 = k$, then the characteristic equation (Eq. (3.98)) becomes:

$$m^2\omega^4 - 4km\omega^2 + 3k^2 = 0, \tag{3.105}$$

from which we find ω^2 to be:

$$\omega^2 = \frac{2km \pm \sqrt{4k^2m^2 - 3k^2m^2}}{m^2} = \frac{k}{m} \text{ or } \frac{3k}{m}, \tag{3.106}$$

which gives the following values for the first and second natural frequencies of the system:

$$\omega_1 = \sqrt{\frac{k}{m}}, \text{ and } \omega_2 = \sqrt{\frac{3k}{m}}. \tag{3.107}$$

By plugging these values for natural frequencies into Eqs. (3.103) and (3.104), we find:

$$r_1 = \left(\frac{Y_1}{Y_2}\right)_1 = \frac{-m\frac{k}{m} + k + k}{k} = 1, \tag{3.108}$$

and

$$r_2 = \left(\frac{Y_1}{Y_2}\right)_2 = \frac{-m\frac{3k}{m} + k + k}{k} = -1, \tag{3.109}$$

which imply that that the first mode of the system is the mode in which $Y_1 = Y_2$, and as a result both masses move together in the same direction and with the same magnitude (Fig. 3.7), and the second mode is the mode in which $Y_1 = -Y_2$, and as a result the two masses move in the opposite directions at all times, but with the same magnitude (Fig. 3.7). The general response of this system will be a combination of the responses that are observed for the two modes:

Fig. 3.7 Modes 1 and 2 of the 2 DOF system of Fig. 3.6

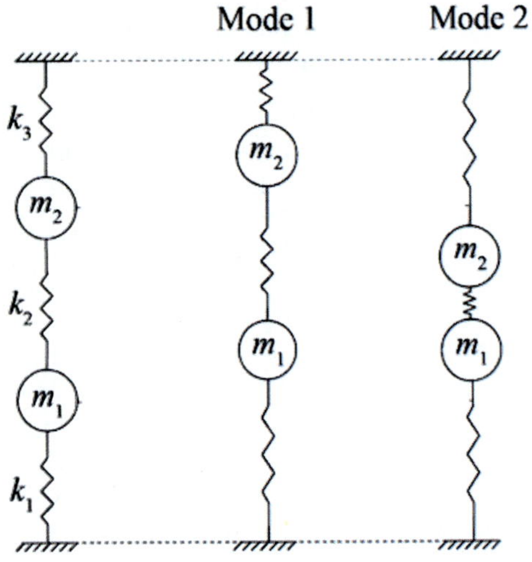

$$
\begin{aligned}
y_1(t) &= Y_1 \sin\left(\sqrt{\frac{k}{m}}t + \phi_1\right) + Y_2 \sin\left(\sqrt{\frac{3k}{m}}t + \phi_2\right) \\
y_2(t) &= Y_1 \sin\left(\sqrt{\frac{k}{m}}t + \phi_1\right) - Y_2 \sin\left(\sqrt{\frac{3k}{m}}t + \phi_2\right)
\end{aligned}
\tag{3.110}
$$

where Y_1, Y_2, ϕ_1, and ϕ_2 will have to be found using the initial conditions of the two masses.

7 Oscillations of a Structure in Otherwise Still Fluid

Here, we consider the case where a 1 DOF mass-spring-damper system that we discussed in Sect. 5 is placed in otherwise still fluid (Fig. 3.8). We assume that the structure has a circular cross section. We need to pick a cross section, because the structure is in contact with fluid and its cross section influences the forces that act from the fluid on it. If we disturb the cylinder, for example, by giving it an initial displacement and letting it go, then it will start oscillating. The amplitude of these oscillations will decrease and reach zero after a while. This is similar to what we had observed before, in a classical vibrations case with a structural damping, but with no

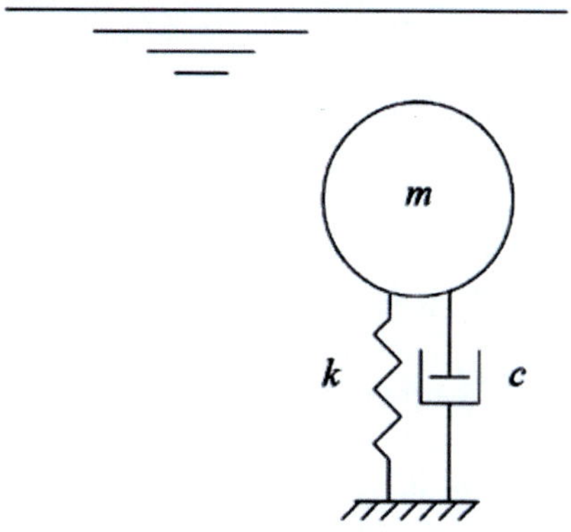

Fig. 3.8 A 1 DOF system in otherwise still fluid

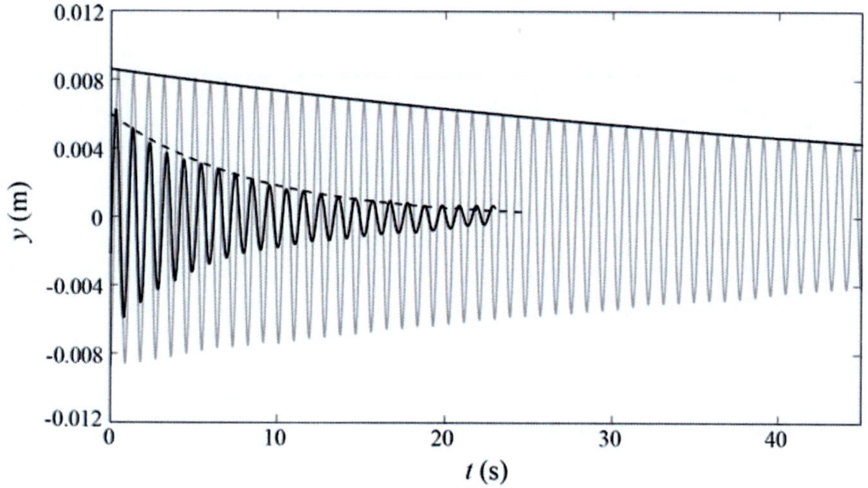

Fig. 3.9 Decaying time histories of displacement of a circular cylinder placed in vacuum (thin line) and in otherwise still fluid (thick line)

fluid around the structure; however, there will be quantitative differences, because now the structure is in contact with fluid. Figure 3.9 shows these differences for a sample case: Oscillations damp out faster when the cylinder is in contact with fluid compared with the case with no fluid around. It is also clear that the frequencies of the decaying oscillations are different in these two cases. Here, we discuss the sources of these quantitative differences.

The drag force that acts on a cylinder as it moves in otherwise still fluid damps the cylinder's oscillations. The drag force is always in the opposite direction of the cylinder's oscillations and dissipates its energy. We call this damping *fluid-induced damping*, which obviously did not exist in the case of a mass-spring-damper in vacuum. We can write the fluid-induced damping due to the drag force as:

$$F_D = \frac{1}{2}\rho C_D D\dot{y}|\dot{y}|, \tag{3.111}$$

where ρ is the fluid density, C_D is the drag coefficient, D is the cylinder's diameter, and $\dot{y}$ is the velocity of the cylinder's oscillations—the only velocity in the system, since the flow velocity is zero. We write $\dot{y}^2$ as $\dot{y}|\dot{y}|$ to ensure that the drag force is always acting in a direction that is opposing the structure's motion.

Now, if we think about the fluid particles that are in contact with the cylinder, initially they are at rest, as everything else in the system. When the cylinder starts to move, then the fluid particles attached to it should move as well. This means that these particles should go from zero velocity to a nonzero velocity over some time, which implies that they will have to be accelerated. These are the particles that are in contact with the structure and now they are accelerating. The fact that they accelerate means that there is a force from the structure acting on them, which is proportional to their acceleration. Then according to Newton's third law, there will be an equal force acting from the fluid on the structure. The structure therefore feels a force proportional to the acceleration from the fluid around it. Since this is a force proportional to the acceleration, it is an inertial force, which can be written as $m_a\ddot{y}$, where m_a is a mass and $\ddot{y}$ is the structure's acceleration. We call this extra mass, which is added to the system due to the fluid, the *added mass*: the mass that is added to the original mass of the structure due to the presence of fluid around it. We can write the inertial force due to the added mass as

$$F_A = m_a\ddot{y}, \tag{3.112}$$

where $m_a = C_m\rho A$ is the added mass, which is written in terms of an added mass coefficient, C_m, the fluid density, ρ, and the area of the displaced fluid, A, which is equal to the cross-sectional area of the structure.

Both the fluid-induced damping force and the added mass force are given per unit length, so that we do not have to worry about the length of the structure for now. This was not an issue before when we did not have any fluid in our system, but it is an issue here, because the fluid force that the structure feels does depend on its length. By taking all the forces per unit length, we do not worry about the third dimension, and the problem stays two-dimensional—in a plane.

Overall, the fluid forces that act on the structure are the added mass force and the fluid-induced damping force, which we can write as:

$$F_{fluid} = -C_m\rho A\ddot{y} - \frac{1}{2}\rho C_D D\dot{y}|\dot{y}|. \tag{3.113}$$

The negative signs are because we write these terms as forces external to the structure. Eq. (3.113) is called the *Morison equation*, and it is the simplest equation to represent the fluid forces that act on a structure oscillating in an otherwise still fluid (Morison et al., 1950).

If we consider this force as an external force acting on a mass-spring-damper:

$$m\ddot{y} + c\dot{y} + ky = F_{fluid}, \tag{3.114}$$

then the equation of motion for our system becomes:

$$m\ddot{y} + c\dot{y} + ky = -\rho C_m A\ddot{y} - \frac{1}{2}\rho C_D D\dot{y}|\dot{y}|. \tag{3.115}$$

Remember that the units for all the terms in this equation are N/m, i.e. for example, m is in kg/m. A major difference between Eq. (3.115) and the general form for a vibratory system with external force (Eq. (3.58)) is that in Eq. (3.115), the right-hand side is also a function of the unknown, y, since flow forces do depend on the motion of the structure.

Since the external force depends on the structure's velocity and acceleration, $\dot{y}$ and $\ddot{y}$, we can move all the terms to the left-hand side to get:

$$(m + \rho C_m A)\ddot{y} + \left(c + \frac{1}{2}\rho C_D D|\dot{y}|\right)\dot{y} + ky = 0. \tag{3.116}$$

This is a nonlinear equation of a self-excited system (i.e., a free vibration case), since there is no external force on the right-hand side anymore. The nonlinearity is due to the term $\dot{y}|\dot{y}|$ in the equation. Also, C_m is not necessarily a constant and could depend on the system's response, which then could add another source of nonlinearity. Dealing with nonlinear equations is not a surprise when we model a system in contact with fluid. The added mass and the fluid-induced damping terms are in fact representing the Navier-Stokes equations, which describe the motion of fluid and are nonlinear equations. To make Eq. (3.116) simpler, we assume that C_m is constant. That means that the added mass will not change with changing the amplitude and frequency of the system's response. This is a major assumption and does not hold true all the time, but it gives us an idea about what the system does. We further make the assumption that we can neglect the dependency of the drag force on the absolute value of the cylinder velocity and we can write it as (Sumer & Fredsoe, 2006):

$$\frac{1}{2}\rho C_D D|\dot{y}|\dot{y} = 2\zeta_f\omega_n(m + m_a)\dot{y}. \tag{3.117}$$

With these assumptions, we can rewrite Eq. (3.116) as:

$$(m + m_a)\ddot{y} + 2(\zeta + \zeta_f)\omega_n(m + m_a)\dot{y} + ky = 0, \tag{3.118}$$

in which

$$\omega_n = \sqrt{\frac{k}{m + m_a}}, \tag{3.119}$$

is the natural frequency of a mass-spring-damper system in otherwise still fluid, ζ is the structural damping ratio, and ζ_f is the fluid-induced damping ratio. The form of this equation is the same as the form of Eq. (3.27), which represented the equation of motion for a 1 DOF classical vibrations system. All we need to do is to find ζ_f and m_a, and then we will have the solution of this system, because its solution follows the general solution of a 1 DOF damped system that we saw in Fig. 3.5. For a structure moving in otherwise still fluid, we can find ζ_f in a similar way that we found ζ for a mass-spring-damper in vacuum, as discussed in Sect. 4 and Fig. 3.3. This time we give an initial displacement to the structure placed in fluid and measure its displacement, which of course decays with time. We can then fit another exponential curve to the decaying signal and find the total damping, which will be $\zeta + \zeta_f$ (see Fig. 3.9). Since we already have ζ, we can find ζ_f from this measurement. We have to be careful: The fluid-induced damping that we calculate here is for a structure placed in otherwise still fluid and with the assumption of Eq. (3.117). As soon as we have flow (nonzero value for incoming flow velocity), the *flow-induced damping* will be different from what we have measured here. And, the values of flow-induced damping can change from one flow velocity to another, as we will discuss later in Chap. 4.

In order to calculate the added mass, m_a, in Eq. (3.118), we use the potential flow assumption. To do so, we need to (i) accelerate the body, (ii) calculate the pressure on the surface of the body, (iii) determine the force on the body by integrating the pressure on the surface, and (iv) write the resulting force as $F = m_a\ddot{y}$ to find m_a.

For a circular cylinder placed in flow, the flow potential is (Fox & McDonald, 1998)

$$\phi = U\left(r + \frac{R^2}{r}\right)\cos\theta. \tag{3.120}$$

Since the cylinder is placed in still fluid, we subtract the flow potential of a constant flow velocity, U, to get

$$\phi = U\frac{R^2}{r}\cos\theta. \tag{3.121}$$

Then, the flow velocities in the radial and circumferential directions can be found as:

$$\begin{aligned} V_r &= \frac{\partial\phi}{\partial r} = -U\frac{R^2}{r^2}\cos\theta, \\ V_\theta &= \frac{1}{r}\frac{\partial\phi}{\partial\theta} = -U\frac{R^2}{r^2}\sin\theta, \end{aligned} \tag{3.122}$$

which give a total flow velocity of

$$V_{Total} = U\frac{R^2}{r^2}, \tag{3.123}$$

which is independent from θ, and becomes $V_{Total} = U$ at the cylinder's surface ($r = R$). Since the flow is assumed to be potential, Bernoulli's equation is valid. Bernoulli's equation for unsteady incompressible flow states that:

$$\rho\frac{\partial\phi}{\partial t} + P + \frac{1}{2}\rho V^2 + \rho gz = \text{constant}, \tag{3.124}$$

where P is the pressure, V is the flow velocity, and z is the elevation. In the present case, z can be assumed to be constant and therefore:

$$\rho\frac{\partial\phi}{\partial t} + P + \frac{1}{2}\rho U^2 = C, \tag{3.125}$$

or

$$P = -\rho\frac{\partial\phi}{\partial t} - \frac{1}{2}\rho U^2 + C, \tag{3.126}$$

where C is a constant. By plugging the flow potential (Eq. (3.121)) into Eq. (3.126), we find:

$$P = -\rho\frac{\partial}{\partial t}\left(U\frac{R^2}{r}\cos\theta\right)_{r=R} - \frac{1}{2}\rho U^2 + C, \tag{3.127}$$

or

$$P = -\rho\dot{U}R\cos\theta - \frac{1}{2}\rho U^2 + C, \tag{3.128}$$

where $\dot{U}$ is the acceleration of the flow in contact with the cylinder, which must be equal to the cylinder acceleration, thus $a = \dot{U} = \ddot{y}$, and Eq. (3.128) is written as:

$$P = -\rho aR\cos\theta - \frac{1}{2}\rho U^2 + C. \tag{3.129}$$

Now we can integrate the pressure over the entire surface to find the flow forces in the direction of motion:

$$F_P = \int_0^{2\pi} PR\cos\theta d\theta. \tag{3.130}$$

We plug Eq. (3.128) into Eq. (3.130):

$$F_P = \int_0^{2\pi} \left(-\rho a R \cos\theta - \frac{1}{2}\rho U^2 + C\right) R \cos\theta d\theta, \tag{3.131}$$

which results in three separate integrals:

$$F_P = \int_0^{2\pi} (-\rho a R \cos\theta) R \cos\theta d\theta + \int_0^{2\pi} \left(-\frac{1}{2}\rho U^2\right) R \cos\theta d\theta + \int_0^{2\pi} C R \cos\theta d\theta, \tag{3.132}$$

or:

$$\begin{aligned} F_P &= \rho a R^2 \int_0^{2\pi} \cos^2\theta d\theta - \frac{1}{2}\rho U^2 R \int_0^{2\pi} \cos\theta d\theta + C R \int_0^{2\pi} \cos\theta d\theta \\ &= -\rho a R^2 \pi + 0 + 0, \end{aligned} \tag{3.133}$$

which gives us:

$$F_P = -\rho a R^2 \pi. \tag{3.134}$$

Then we rewrite F_P as the added mass times acceleration:

$$F_P = -\rho \pi a R^2 \equiv -m_a a, \tag{3.135}$$

which results in the following equation for the added mass:

$$m_a = \rho \pi R^2. \tag{3.136}$$

The negative sign in Eq. (3.135) is because the force that we find here is the force acting from the fluid on the structure.

Following Eq. (3.136), the added mass coefficient for a circular cylinder placed in otherwise still fluid based on the potential flow assumption is $C_m = 1$. Later on (in Chap. 4), we will see how this coefficient changes as the cylinder is placed in fluid flow (as opposed to still fluid).

After knowing the values of m_a and ζ_f, the solution of Eq. (3.118) is the same as that of the classical 1D vibrations system, which we discussed in Sect. 3. The amplitude and frequency plots of Fig. 3.5 will be valid for this case as well, when the natural frequency, as defined in Eq. (3.119), and the total damping ratio, $\zeta + \zeta_f$, are considered. All the fluid does in this case is to change the system's natural frequency and damping ratio and increase the damping rate of the structure's response. The fluid does not make the structure oscillate. The structure starts to oscillate first, and then the fluid moves around it. In the following chapter, we will consider a case where the structure oscillates due to flow forces.

Chapter 4
Vortex-Induced Vibrations

1 Introduction

In this chapter, we consider one of the fundamental problems in FSI: that of a mass-spring-damper placed in flow. Although this is a rather simple system, it results in rather complicated interactions between the flow and the structure, and it helps us understand some fundamental concepts in FSI. The vibrations that we will observe in this system are due to the shedding of vortices in the wake of the structure, and that is why we call this problem vortex-induced vibrations (VIV). In this chapter, we consider a body with a circular cross-section. Several studies exist on VIV of a flexibly mounted cylinder, and here we will go over some fundamentals that we need to understand it. Three review papers on this subject that are widely cited are Bearman (1984), Sarpkaya (2004), and Williamson and Govardhan (2004). First, we consider a cylinder which is allowed to move in one direction only. To begin with, we consider the case where the degree of freedom is perpendicular to the direction of the flow, i.e., crossflow direction (CF), and we will study several different concepts in crossflow VIV. Then, we will consider the case where the degree of freedom is in the direction of the flow, i.e., the inline direction (IL). At the end of the chapter, very briefly, we will consider a cylinder, which is allowed to oscillate in both the CF and the IL directions, i.e., a 2 DOF system.

2 Vortex-Induced Vibrations of a One-Degree-of-Freedom System

Consider a cylinder attached to a spring and a damper, and placed in fluid flow (Fig. 4.1). This is in fact a classical vibrations problem of a mass-spring-damper, which we discussed in Chap. 3, but subjected to fluid flow. Here, we focus on a case

Y. Modarres-Sadeghi, *Introduction to Fluid-Structure Interactions*,
https://doi.org/10.1007/978-3-030-85884-1_4

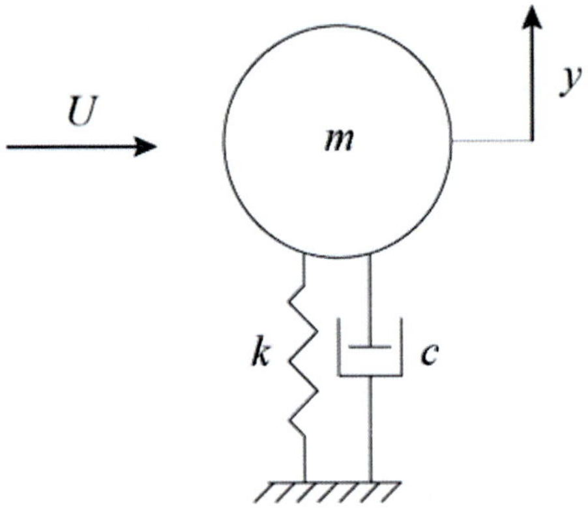

Fig. 4.1 A flexibly mounted cylinder placed in flow with one degree of freedom in the CF direction (y-direction, or the direction perpendicular to the flow)

where the structure can oscillate only in one direction and that direction is perpendicular to the direction of flow, i.e., the transverse direction (y-direction in Fig. 4.1). This is what we call the crossflow (CF) direction.

As we discussed in Chap. 2, due to the presence of a bluff body (a cylinder in this case) in flow, vortices are shed downstream of the structure, which result in fluctuating lift and drag forces acting on the structure. If the structure is free to oscillate, which is the case in the problem we are discussing here, flow forces can potentially make the structure oscillate. The structure oscillates only if the frequency of vortex shedding is close to the natural frequency of the structure. This is similar to what we call resonance in vibrations of a mass-spring-damper as we discussed in Chap. 3: the structure oscillates when the frequency of external force is equal to the natural frequency of the system. The frequency of external force in our problem here is the shedding frequency. One major difference between our system and a classical vibrations system is that our system here is the mass-spring-damper plus the flow, i.e., the structure and the flow together constitute our system. Therefore, the natural frequency of our system here as a whole can change with the flow velocity. Since the cylinder is in contact with flow, when it wants to move "naturally" at a frequency equal to its natural frequency, it has to push the flow around it away. Therefore, the structure feels an added mass as we discussed in Chap. 3 which does influence the natural frequency. We have to be careful here. What is being pushed away by the structure is the fluid flow, not the still fluid. This means that the added mass will not necessarily be equal to the added mass that we discussed in Chap. 3 for a cylinder moving in otherwise still fluid. Also it means that the added mass will not be constant at different flow velocities. This is different from what we had seen before in a classical vibrations system. In that case, we could change the frequency of external force to hit the resonance and get large amplitude oscillations. But here, in order to change the frequency of external force, we have to change the flow velocity, which means that we are changing the added mass, and thus the natural frequency as well.

3 The Equation of Motion and the Major Dimensionless Parameters

For our 1 DOF VIV system, the equation of motion is:

$$m\ddot{y} + c\dot{y} + ky = F_y, \tag{4.1}$$

in which m, c, and k are the mass, damping, and stiffness of the structure only (without any influence of the flow), and F_y includes all the flow forces in the transverse direction. Since the cylinder has a length and we would like to avoid dealing with it in our analysis of the system, we consider all the parameters in this equation, i.e., m, c, k, and F_y, as per unit length parameters. Then the unit for m is kg/m and for F_y is N/m. The mean value of F_y is zero, due to the symmetry of the system, as long as we consider a circular cross-section. The fluctuating component of this force is nonzero due to the vortex shedding and we can write it as:

$$F_y = \frac{1}{2}\rho D U^2 c_y \sin(\omega_s t + \phi), \tag{4.2}$$

in which ω_s is the frequency of vortex shedding in rad/s, $\frac{1}{2}\rho D U^2 c_y$ is the magnitude of the fluctuating force in the y-direction (this form has been used extensively for the magnitude of flow forces, e.g., in representing the lift and drag forces that act on fixed bodies, as we discussed in Chap. 2), and ϕ is the phase difference between the fluctuating force and the cylinder's displacement. In the magnitude of force, the fluid density, ρ, and the cylinder diameter, D, are constants for a given fluid and a given cylinder, and the flow velocity, U, can change. As a result of changing the flow velocity, the flow forces change and that is reflected in c_y, which is the coefficient of flow force in the y-direction. We can write Eq. (4.1) in terms of the structure's natural frequency and damping ratio as

$$\ddot{y} + 2\zeta\omega_n\dot{y} + \omega_n^2 y = \frac{1}{2}\frac{\rho}{m} D U^2 c_y \sin(\omega_s t + \phi). \tag{4.3}$$

Note that here we have all the flow forces on the right-hand side, which means that the natural frequency that we see on the left-hand side of Eq. (4.3) is the natural frequency of the structure in vacuum, with no fluid or flow around it, and the damping is the pure structural damping.

Before we do much with this equation, we make it dimensionless. Through this process, we will see the major dimensionless parameters that rule our system. The first step toward making this equation dimensionless is to define some dimensionless parameters. Since we expect to see some oscillations in the system, the amplitude and frequency of oscillations are the important parameters for us. For the amplitude, we make it dimensionless by dividing it by the cylinder's diameter:

$$y^* = \frac{y}{D}, \tag{4.4}$$

where y^* is the dimensionless amplitude. For the frequency, we make it dimensionless using the cylinder diameter, D, and the flow velocity, U, and we write it in terms of a dimensionless time:

$$t^* = \frac{tU}{D}, \tag{4.5}$$

where t^* is the dimensionless time and t is the dimensional time. Note that this is very similar to the inverse of the Strouhal number as we discussed in Chap. 2.

Now, let's use these two dimensionless parameters and make our equation of motion dimensionless. We first write Eq. (4.3) as:

$$\frac{\mathrm{d}^2 y}{\mathrm{d}t^2} + 2\zeta\omega_n \frac{\mathrm{d}y}{\mathrm{d}t} + \omega_n^2 y = \frac{1}{2} c_y \frac{\rho D}{m} U^2 \sin(\omega_s t + \phi), \tag{4.6}$$

and then plug relations (4.4) and (4.5) into it:

$$\frac{\mathrm{d}^2(y^* D)}{\mathrm{d}\left(\frac{t^* D}{U}\right)^2} + 2\zeta\omega_n \frac{\mathrm{d}(y^* D)}{\mathrm{d}\left(\frac{t^* D}{U}\right)} + \omega_n^2 y^* D = \frac{1}{2} c_y \frac{\rho D}{m} U^2 \sin\left(\omega_s \frac{t^* D}{U} + \phi\right). \tag{4.7}$$

Then we bring the constants out and simplify it to:

$$\frac{U^2}{D} \frac{\mathrm{d}^2 y^*}{\mathrm{d}t^{*2}} + 2\zeta\omega_n U \frac{\mathrm{d}y^*}{\mathrm{d}t^*} + \omega_n^2 y^* D = \frac{1}{2} c_y \frac{\rho D}{m} U^2 \sin\left(\omega_s \frac{t^* D}{U} + \phi\right), \tag{4.8}$$

and we multiply both sides by D/U^2, multiply the right-hand side by $(\pi/4)/(\pi/4)$, use $\omega = 2\pi f$, and organize everything to get:

$$\begin{aligned} &\frac{\mathrm{d}^2 y^*}{\mathrm{d}t^{*2}} + (4\pi\zeta) \frac{f_n D}{U} \frac{\mathrm{d}y^*}{\mathrm{d}t^*} + (2\pi)^2 y^* \frac{(f_n D)^2}{U^2} \\ &\quad = \frac{1}{2} c_y \frac{1}{\pi/4} \left(\frac{(\pi/4)\rho D^2}{m} \right) \sin\left(2\pi t^* \frac{f_s D}{U} + \phi\right). \end{aligned} \tag{4.9}$$

This is the dimensionless form of the equation of motion for a flexibly mounted cylinder placed in flow. In this dimensionless form of the equation, the following two terms show up:

$$\frac{f_n D}{U} \equiv \frac{1}{V_r}, \tag{4.10}$$

and

$$\frac{(\pi/4)\rho D^2}{m} \equiv \frac{1}{m^*}. \tag{4.11}$$

The inverse of the first term is a dimensionless flow velocity, which depends both on the flow properties, U, and the structural properties, f_n and D. We call this dimensionless velocity, *the reduced velocity*:

$$V_r \equiv \frac{U}{f_n D}. \tag{4.12}$$

The inverse of the second term is a dimensionless ratio between the mass of the structure and the mass of displaced fluid. We call this dimensionless number, *the mass ratio*:

$$m^* \equiv \frac{m}{(\pi/4)\rho D^2}. \tag{4.13}$$

Then we can re-write Eq. (4.9) with these two parameters as

$$y^{*\prime\prime} + \frac{4\pi\zeta}{V_r} y^{*\prime} + \left(\frac{2\pi}{V_r}\right)^2 y^* = \frac{2}{\pi} c_y \frac{1}{m^*} \sin\left(2\pi \frac{f_s D}{U} t^* + \phi\right), \tag{4.14}$$

where $(\ )'$ denotes the derivate with respect to the dimensionless time: $d()/dt^*$. After multiplying both sides by m^*, we will get:

$$m^* y^{*\prime\prime} + \frac{4\pi\zeta m^*}{V_r} y^{*\prime} + \left(\frac{2\pi}{V_r}\right)^2 m^* y^* = \frac{2}{\pi} c_y \sin\left(2\pi \frac{f_s D}{U} t^* + \phi\right). \tag{4.15}$$

Since we know the reduced velocity, V_r, has the system's natural frequency, f_n, in its denominator, then we call its inverse, a reduced frequency, f_r:

$$f_r = \frac{1}{V_r}. \tag{4.16}$$

Also, on the right-hand side of Eq. (4.15), we have a dimensionless frequency, $f_s D/U$, which we call f_{rs}, and then rewrite Eq. (4.15) in terms of f_r and f_{rs}:

$$m^* y^{*\prime\prime} + 2\zeta m^* (2\pi f_r) y^{*\prime} + (2\pi f_r)^2 m^* y^* = \frac{2}{\pi} c_y \sin(2\pi f_{rs} t^* + \phi), \tag{4.17}$$

and then since $2\pi f_r \equiv \omega_r$ and $2\pi f_{rs} \equiv \omega_{rs}$, rewrite it as

$$m^* y^{*\prime\prime} + 2\zeta m^* \omega_r y^{*\prime} + \omega_r^2 m^* y^* = \frac{2}{\pi} c_y \sin(\omega_{rs} t^* + \phi). \tag{4.18}$$

If we compare this equation with the equation for a classical vibrations system with an external sinusoidal force:

$$m\ddot{y} + 2\zeta m\omega_n \dot{y} + \omega_n^2 my = F_0 \sin \omega t, \tag{4.19}$$

we see that they have similar forms: m^* has the same role as m; ζ is the same in both equations; ω_r has the same role as ω_n, and ω_s as ω. This similarity makes it easier to relate what we will observe in the VIV response of a cylinder to what we know from the classical vibrations problem of a mass-spring-damper system. Despite the similar format, there is a major difference between the two systems: ω_r depends on flow forces, and changes with flow velocity for a given VIV system, while ω_n is a constant value for a classical vibrations system. This difference is the source of several new phenomena that we observe in VIV compared with a classical vibrations system. Without this difference, VIV would have been a boring extension of the classical vibrations. But the fun begins here.

4 Amplitude and Frequency Responses

Here, we look at the cylinder's response as the flow velocity is increased. This is shown in Fig. 4.2 based on experimental results for a cylinder with $m^* = 4.6$ and $\zeta = 0.01$. We should know these two parameters to know our system. They show up on the left-hand side of Eq. (4.18). The third parameter that we need to know is the reduced velocity, which also shows up in Eq. (4.18) as $\omega_r = 2\pi f_r = 2\pi / V_r$. The reduced velocity, $V_r = U/(f_n D)$, varies as we change the flow velocity. We could have varied the reduced velocity by varying D or f_n, but here we do it by varying the flow velocity, and we keep D and f_n constant throughout the experiments.

In Fig. 4.2, the dimensionless amplitude of response, defined as $A^* = A/D$, and the dimensionless frequency of response, defined as $f^* = f/f_n$, are shown as the reduced velocity, V_r, is varied. The straight line in the frequency plot shows the expected shedding frequency off a fixed cylinder with a Strouhal number of $St = 0.2$. As the shedding frequency increases with increasing reduced velocity, at some point, the shedding frequency equals the natural frequency of the system. This coalescence of the forcing frequency and the natural frequency makes the structure oscillate, which is similar to resonance in a classical vibrations example. In the amplitude plot of Fig. 4.2, these oscillations start at $V_r \sim 4.0$ (point A in the figure). As soon as the structure starts to oscillate, the shedding frequency follows the oscillation frequency, instead of following the Strouhal line, as if the cylinder pushes the flow back and forth and controls the shedding of vortices. As we increase the reduced velocity further, the amplitude of oscillations increases initially to values of $A^* \sim 0.3$ (point B) and then it jumps to higher values of $A^* \sim 0.7$ at a reduced velocity of $V_r \sim 5.4$

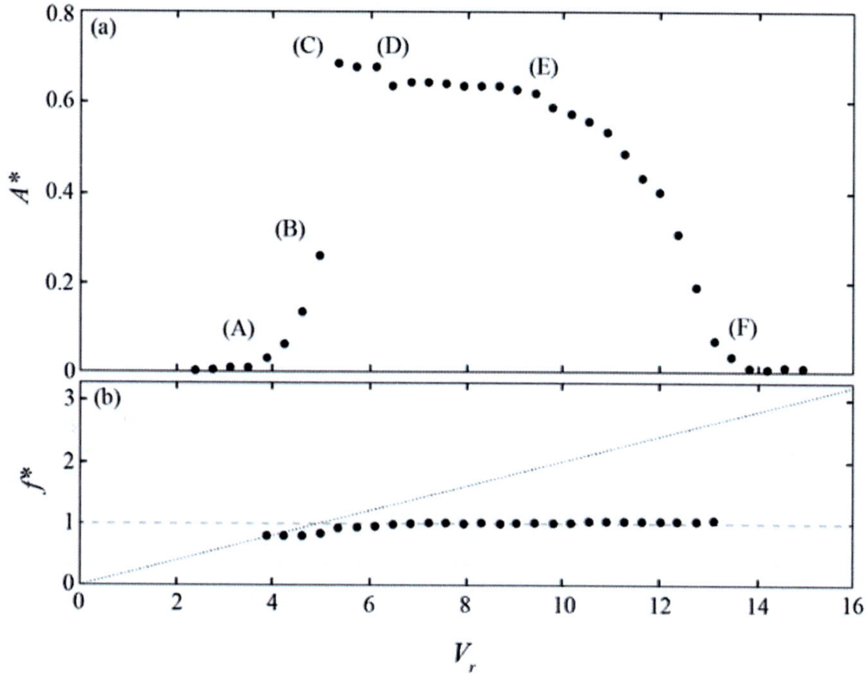

Fig. 4.2 Dimensionless (**a**) amplitude and (**b**) frequency of oscillations versus the reduced velocity for a cylinder with one degree of freedom in the CF direction placed in flow

(point C). Then the amplitude stays at $A^* \sim 0.7$ for a short range of reduced velocities up to $V_r \sim 6$, where a sudden decrease in the amplitude is observed (point D). For higher reduced velocities, the amplitude stays almost constant up to $V_r \sim 9.4$ (point E) and then decreases monotonically until it goes back to zero at $V_r \sim 12.8$ (point F). This is the typical behavior of a 1 DOF cylinder placed in flow. The AB branch is called the *initial branch*, the CD branch the *upper branch*, and the DF branch the *lower branch*. In the range of nonzero amplitude, the oscillation frequency remains close to the system's natural frequency. This is the range in which the cylinder's oscillation rules the shedding of vortices, and as a result, the shedding frequency is equal to the oscillation frequency. This shedding frequency is the shedding frequency off the oscillating cylinder, not a fixed cylinder, which deviates from the frequency line predicted based on the Strouhal law.

The range of reduced velocities for which a non-zero amplitude is observed and for which the frequency of oscillations, the shedding frequency, and the natural frequency of the system are all the same is called the *lock-in range*. The maximum amplitude of oscillations for the case of Fig. 4.2 is around $A/D = 0.7$. Typically, amplitudes of up to around one cylinder diameter are observed in the VIV response of cylinders with one degree of freedom in the crossflow direction. Figure 4.3 is a schematic of a typical VIV response with the three branches that we discussed. In cases with large mass ratio and large damping ratio, the upper branch does not appear

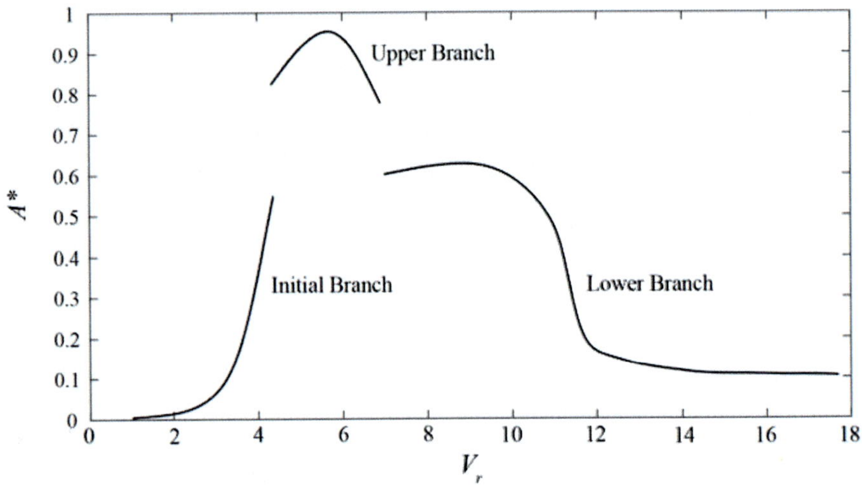

Fig. 4.3 Schematic of a typical amplitude response for a system with a low mass ratio and a low damping ratio with an upper branch in the response. (Reproduced from Williamson & Govardhan, 2004)

as an independent branch, but rather as the beginning of the lower branch, and we see two branches only: initial and lower. The upper branch is observed for light and lightly damped systems only.

5 The Influence of Mass and Damping on the VIV Response

If we keep everything else constant, and increase the cylinder's mass, i.e., a larger m^*, the maximum amplitude of oscillations decreases. Also, if we keep everything else constant, and increase the structural damping, ζ, again the maximum amplitude of oscillations decreases. Then we can say that the maximum amplitude of oscillations decreases by increasing $m^*\zeta$. We call $m^*\zeta$, *the mass-damping parameter.*

The plot of Fig. 4.2 is for a case with a mass ratio of $m^* = 4.6$ and a damping ratio of $\zeta = 0.01$, resulting in a mass-damping parameter of $m^*\zeta = 0.046$. A mass-damping parameter of $m^*\zeta = 0.046$ is considered a low mass-damping parameter, a mass ratio of $m^* = 4.6$ is considered a low mass ratio, and a damping ratio of $\zeta = 0.01$ is considered a low damping ratio too. This mass ratio is considered a small value compared with a case where the cylinder is placed in air flow instead of water flow. Since the air density is much lower than the water density, then the mass ratio of a system in air is much larger than the mass ratio of the same system in water, which leads to a larger mass-damping parameter. We do not have to go from water to air to change the mass-damping parameter. We can stay in water and use a much heavier cylinder to increase the mass ratio, or use roller bearings instead of air bearings to increase the damping ratio, and thus get a larger mass-damping

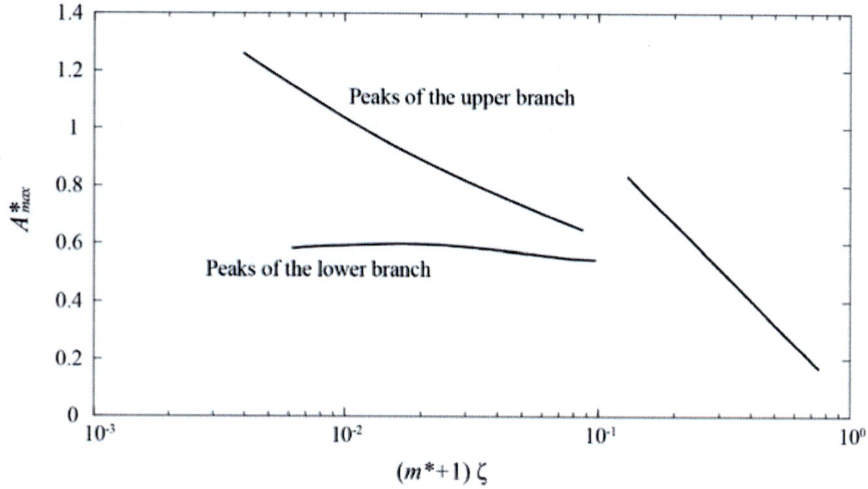

Fig. 4.4 Schematic of how the peak amplitude changes with the mass-damping parameter (Reproduced from Williamson & Govardhan, 2004)

parameter. If the mass-damping parameter is very small, then the upper branch can have much larger amplitudes (around one diameter).

Figure 4.4 gives a general view of how the mass-damping parameter influences the peak amplitude observed in the VIV response. The horizontal axis is not exactly the mass-damping parameter: it is $(m^* + 1)\zeta$ instead of $m^*\zeta$, but it still represents the same effect. The large mass-damping parameters are based on tests in air and the low mass-damping parameters are based on tests in water. Two peaks are shown for the low mass-damping parameter cases, one of which corresponds to the maximum amplitude observed in the lower branch and the other one to the maximum amplitude observed in the upper branch.

6 Shedding Patterns at Different Reduced Velocities

In this section, we discuss the wake of a cylinder as it undergoes VIV. In Chap. 2 and for a fixed cylinder, we observed a von Kármán Street in the wake for $Re > 47$. When a von Kármán Street is observed, one vortex is shed from each side of the cylinder and a shedding frequency is established based on this shedding pattern. If the cylinder starts to oscillate in the crossflow direction, which is what we observe in VIV, initially when the amplitude of oscillations is rather small, we do not expect to see a major difference in the shedding pattern compared with a rigid cylinder: For low reduced velocities, where we observe the initial branch response, a shedding pattern very similar to a classical von Kármán Street is observed, in which one vortex is shed from each side of the cylinder during one cycle of oscillations. We call this a *2S pattern*, for 2 *single* vortices that are shed during each cycle. Figure 4.5a shows a

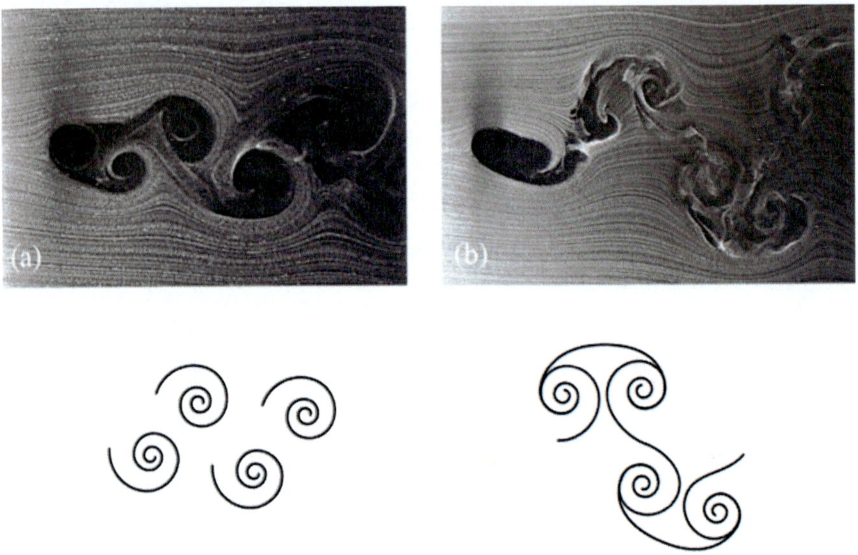

Fig. 4.5 Samples of (**a**) 2S and (**b**) 2P vortex patterns in the wake of a cylinder

snapshot of flow visualization of the 2S pattern in the wake of a cylinder undergoing VIV, together with a schematic.

At higher reduced velocities, where we observe the lower branch response, instead of two single vortices in the wake of the cylinder during each cycle of oscillations, we observe two pairs of vortices. In this case, as a shear layer rolls from one side and forms a vortex, the cylinder moves and brings the vortex close to the shear layer that has just left the other side of the cylinder. This shear layer then gets inside the vortex and creates a second vortex with an opposite direction of rotation inside the first vortex. This mechanism creates what we call a vortex pair, and since it is repeated twice in each cycle—once when the cylinder moves in one direction, and once when it moves in the other direction, we observe two *pairs* of vortices that are shed in each cycle, and as a result we call this pattern a *2P pattern*. A snapshot of a sample 2P shedding is shown in Fig. 4.5b together with a schematic. The 2P shedding is observed at larger reduced velocities and lower reduced frequencies, where the vortices have more time to interact in the wake and create 2P patterns.

7 The Added Mass and the Flow-Induced Damping

In this section, we discuss how the added mass and the flow-induced damping change in a system undergoing VIV. The equation of motion for a 1D VIV system is (Eq. 4.1):

$$m\ddot{y} + c\dot{y} + ky = F_y, \tag{4.20}$$

in which

$$F_y = \frac{1}{2}\rho D U^2 c_y \sin(\omega t + \phi) = \overline{F}_y \sin(\omega t + \phi), \tag{4.21}$$

where ω is the frequency of the external force that is being applied on the cylinder. In a self-excited VIV case, $\omega = \omega_s$. We use $\overline{F}_y$ for the magnitude of this force to simplify our future discussions. Let's assume that the solution to this equation is a pure sinusoidal function:

$$y(t) = A \sin \omega t. \tag{4.22}$$

We should remember that this is a correct assumption for a linear system. We know, however, that we do not deal with a linear system here, since c_y is not linear, as it is technically representing the Navier-Stokes equations, which are by no means linear. But for now, let's make the assumption that a sinusoidal solution is good enough, which is a widely used assumption. In that case, we can find the velocity and acceleration of the solution as:

$$\dot{y}(t) = A\omega \cos \omega t, \tag{4.23}$$

and

$$\ddot{y}(t) = -A\omega^2 \sin \omega t. \tag{4.24}$$

Then we can expand the sine term in the flow force of Eq. (4.21) as

$$\begin{aligned} F_y(t) =& \overline{F}_y \sin(\omega t + \phi) \\ =& \overline{F}_y(\sin \omega t \cos \phi + \cos \omega t \sin \phi), \end{aligned} \tag{4.25}$$

which includes sine and cosine of ωt, which we can write in terms of velocity and acceleration of the solution using Eqs. (4.23) and (4.24):

$$\dot{y}(t) = A\omega \cos \omega t \quad \Rightarrow \quad \cos \omega t = \frac{\dot{y}(t)}{A\omega}, \tag{4.26}$$

$$\ddot{y}(t) = -A\omega^2 \sin \omega t \quad \Rightarrow \quad \sin \omega t = -\frac{\ddot{y}(t)}{A\omega^2}, \tag{4.27}$$

and then we can plug these into Eq. (4.25) to find

$$F_y(t) = \overline{F}_y\left(\left(-\frac{\ddot{y}(t)}{A\omega^2}\right)\cos \phi + \left(\frac{\dot{y}(t)}{A\omega}\right)\sin \phi\right), \tag{4.28}$$

and rewrite it as:

$$F_y(t) = -\frac{\overline{F}_y \cos\phi}{A\omega^2}\ddot{y}(t) + \frac{\overline{F}_y \sin\phi}{A\omega}\dot{y}(t). \tag{4.29}$$

This is the flow force in terms of a component in phase with acceleration (the first term) and a component in phase with velocity (the second term). The component in phase with acceleration is an inertia force:

$$F_{La} = -\frac{\overline{F}_y \cos\phi}{A\omega^2}\ddot{y}(t), \tag{4.30}$$

which can be written in terms of a mass, multiplied by the cylinder's acceleration:

$$F_{La} = -m_a\ddot{y}(t), \tag{4.31}$$

and the component in phase with velocity is a damping force:

$$F_{Lv} = \frac{\overline{F}_y \sin\phi}{A\omega}\dot{y}(t) \equiv \frac{\frac{1}{2}\rho D U^2 c_y \sin\phi}{A\omega}\dot{y}(t), \tag{4.32}$$

where we can define a new coefficient, C_{LV}, as

$$C_{LV} = \frac{c_y \sin\phi}{A\omega}, \tag{4.33}$$

and then rewrite the flow force in phase with velocity as

$$F_{Lv} = \frac{1}{2}\rho D U^2 C_{LV}\dot{y}(t), \tag{4.34}$$

or in terms of a damping coefficient, multiplied by the cylinder's velocity

$$F_{Lv} = c_f\dot{y}(t), \tag{4.35}$$

where c_f is the flow-induced damping coefficient.

Then the equation of motion (Eq. (4.20)) can be written as:

$$m\ddot{y} + c\dot{y} + ky = -m_a\ddot{y} + \frac{1}{2}\rho D U^2 C_{LV}\dot{y}, \tag{4.36}$$

and by bringing all the terms to the left-hand side:

$$(m + m_a)\ddot{y} + \left(c - \frac{1}{2}\rho D U^2 C_{LV}\right)\dot{y} + ky = 0. \tag{4.37}$$

This form of the equation of motion represents a free-vibration system, since there is no external force on the right-hand side. The flow forces show up as new terms on the left-hand side: m_a is the mass that the system feels due to the flow around it, i.e., the added mass, and $c_f \equiv \frac{1}{2}\rho DU^2 C_{LV}\dot{y}$ is the damping that the system feels due to the flow around it, i.e., the flow-induced damping. In this new form, the system's natural frequency is:

$$\omega_n = \sqrt{\frac{k}{m + m_a}}, \tag{4.38}$$

where m_a can be written in terms of the volume of displaced fluid:

$$m_a = C_m \rho \frac{\pi D^2}{4}, \tag{4.39}$$

in which C_m is the added mass coefficient.

If we know C_m and C_{LV}, then we can solve Eq. (4.37), but C_m and C_{LV} are both functions of the amplitude and frequency of response. Then we need to know the solution to know these parameters, which suggests an iterative way of solving the equation. We will discuss how we can solve this equation in Sects. 8 and 9, but before that, we look at some experimental results for C_m and C_{LV}.

Figure 4.6 shows how C_m and C_{LV} vary with V_r within the lock-in range for the cylinder whose response was shown in Fig. 4.2. The added mass coefficient is equal to 1 for a cylinder moving in otherwise still fluid according to the potential flow assumption, as we discussed in Chap. 2 (dashed line in Fig. 4.6a); however, the added mass does change with changing reduced velocity in a system undergoing VIV. At the beginning of the lock-in range, when the cylinder oscillates with very small amplitudes, the added mass coefficient is close to 1 (slightly smaller). As the reduced velocity is increased, the added mass decreases and approaches values of around zero and even negative values. A negative added mass should not be surprising, because the added mass in fact represents a portion of the flow force that is in phase with acceleration. There is nothing against this portion of the flow force being in the direction opposing the direction of positive acceleration. In Sect. 8, we will discuss the negative added mass values a bit more.

In Eq. (4.37), the sign of the damping coefficient depends on the sign of C_{LV}, since all the other parameters in $\frac{1}{2}\rho DU^2 C_{LV}$ are always positive. If C_{LV} is negative, then the flow-induced damping is positive and it acts in the same way as the structural damping, and the structure's response damps out and reaches zero after a transient. If, however, C_{LV} is positive, then the flow-induced damping term will be a negative value. A negative value for damping does the opposite of what we expect from a damping term: it amplifies the motion. If we give an initial disturbance to a system with a negative damping, then the amplitude of the resulting oscillations increases with time. In a linear system, the amplitude will increase to infinity. In a nonlinear system, such as the system we are considering here, the oscillations will

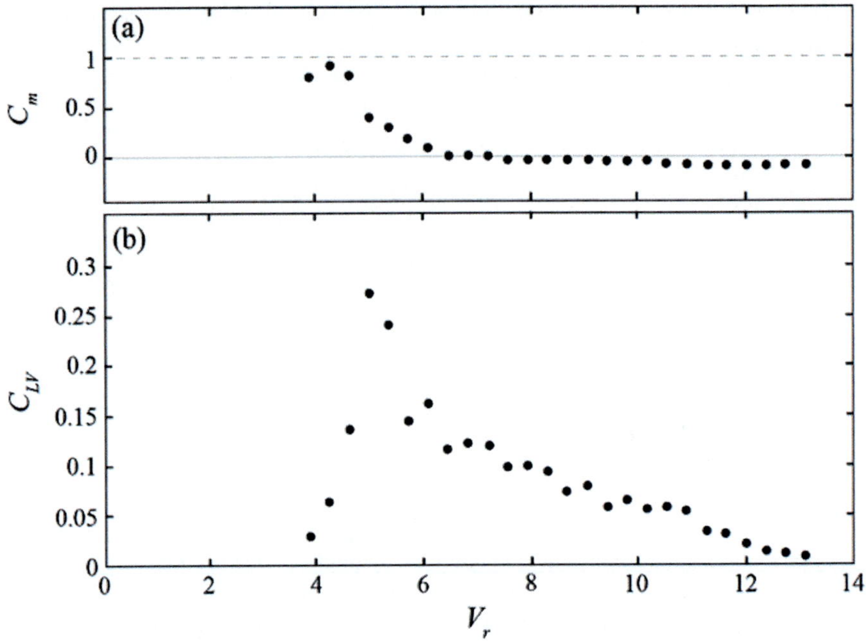

Fig. 4.6 Plots of (**a**) C_m and (**b**) C_{LV} versus V_r for the system of Fig. 4.2

reach a constant magnitude. Here, since there is also a structural damping, which is always positive, then $c - \frac{1}{2}\rho DU^2 C_{LV}$ must be a negative value in order for the system to oscillate, which could be the case only if C_{LV} is positive. Clearly a positive C_{LV} is observed in Fig. 4.6b for the entire lock-in range, although its absolute value changes with V_r.

8 Forced Oscillations of a Cylinder in Flow

Imagine we know the values of C_m and C_{LV} for every possible amplitude and frequency of oscillations. In that case, we can use an iterative method and solve Eq. (4.37). Based on this idea, we can run a series of experiments in which we force a cylinder to undergo a sinusoidal motion at a prescribed frequency and a prescribed amplitude. Then we can measure the flow forces acting on the cylinder and rewrite them in the form of C_{LV} and C_m. Once we have several combinations of different amplitudes and frequencies, we can create databases of C_{LV} and C_m values versus the amplitude and frequency of oscillations. We can then use these databases and solve the equation of motion for a VIV system using iterative methods. This procedure is based on the assumptions that (i) the sinusoidal motions are acceptable representations of the VIV response and (ii) the VIV response does not need any information from its past (i.e., at any instant of oscillations, the flow forces are the same as the flow forces that would have acted on a cylinder oscillating with that amplitude and that frequency permanently). These are not necessarily very correct assumptions, but in order to use the force databases in iterative solutions, we should be aware of these assumptions. Several studies exist on validating these assumptions and on the methods to use the force databases to predict the VIV response. Sarpkaya (2004) has discussed the use of forced databases in his review paper extensively. Gopalkrishnan (1993) has provided a comprehensive set of such databases, as we discuss in what follows.

Figure 4.7a shows a sample of a C_{LV} database based on a series of experimental results conducted by Gopalkrishnan (1993) at a constant Reynolds number of approximately 10,000. This is a 3D plot showing the C_{LV} values versus the imposed amplitude of oscillations and imposed frequency of oscillations. In the plot, the frequency is converted into a dimensionless number similar to the reduced velocity,

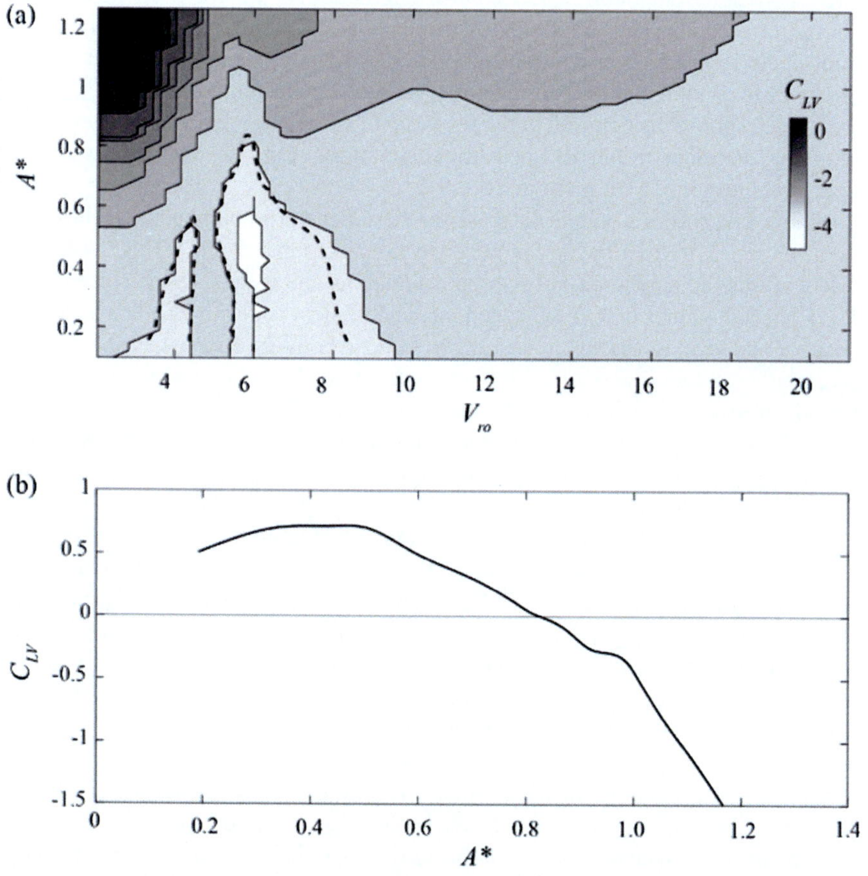

Fig. 4.7 Lift coefficient in phase with velocity, C_{LV}, (**a**) versus the amplitude of imposed oscillations and V_{ro}, and (**b**) versus the amplitude of imposed oscillations for a constant V_{ro} of 6. The dashed lines represent zero. (Reproduced from Gopalkrishnan, 1993)

$V_{ro} = U/f_oD$, but based on the oscillation frequency, f_o. Regions of positive and negative C_{LV} exist at different amplitudes and frequencies of oscillations. Two regions of positive C_{LV} are observed around $V_{ro} = 4$ and $V_{ro} = 6$–10, which suggest oscillations could occur in the system within these regions. Figure 4.7b is a slice of the 3D plot of Fig. 4.7a, which shows the C_{LV} values versus the amplitude of oscillations for a constant V_{ro} of 6. Consider an iterative method to solve the equation of motion using this database. For small amplitudes, C_{LV} is positive, which could potentially lead to a negative damping (with a negligible structural damping), and the excitation of the structure, thus an increase in the amplitude. For larger amplitudes of oscillations, C_{LV} becomes negative, which will result in an overall positive damping in the system, thus reducing the amplitude of oscillations. With smaller amplitudes, we will go back to a range of positive C_{LV} and the resulting increase in the amplitude, and the loop goes on and on. Besides making an iterative solution possible, this loop gives us a feeling for the *limit cycle* nature of VIV, where larger amplitudes are damped and smaller amplitudes are amplified until the system converges to a constant amplitude of oscillations: the limit cycle amplitude. The limit cycle response is a constant amplitude, single-frequency oscillation that is observed in a nonlinear system, where several different initial conditions result in the same response, as opposed to the response of a linear system where any new set of initial conditions results in a new amplitude of the response. We will discuss the limit cycle response a bit more in Sect. 9.

Figure 4.8a shows a sample database for the added mass coefficient, C_m, based on the same series of experiments conducted by Gopalkrishnan (1993). This is a 3D plot of C_m versus the amplitude of imposed oscillations and its frequency, converted to V_{ro}. For small values of V_{ro}, the added mass coefficient is positive and close to 1. A positive added mass means a seemingly heavier body in flow when the body is accelerating and a value of $C_m = 1$ is what the potential flow assumption predicts for the added mass coefficient of a cylinder in otherwise still fluid as we discussed in Chap. 2. As V_{ro} is increased, the added mass coefficient increases initially, which means that the system feels heavier. For larger values of V_{ro}, higher than $V_{ro} \sim 6$, the added mass coefficient drops and goes to negative values. A negative added mass is in fact a "subtracted" mass, which implies that the cylinder feels lighter than it was initially before it was placed in the flow. Within this range, the natural frequency of the system increases compared with the original natural frequency in still fluid, since the added mass is a negative number and reduces the total value in the denominator of Eq. (4.38). This trend for C_m is consistent for different amplitudes of oscillations in Fig. 4.8a. A slice of the 3D plot of Fig. 4.8a for small V_{ro} values is given in Fig. 4.8b to show the trend more clearly.

The major variations in C_m and C_{LV} are observed for $V_{ro} < 10$. Figure 4.9 shows a map of the corresponding shedding patterns for this range of V_{ro} and based on forced VIV tests for amplitudes up to 1.8 times the cylinder diameter. The 2S and 2P patterns that we discussed before in the self-excited VIV response are observed in two regions here. Besides those, an asymmetric S + P pattern is observed for very large amplitudes of prescribed oscillations. This is an example of a case where a rather large amplitude of oscillations gives the vortices enough time to create

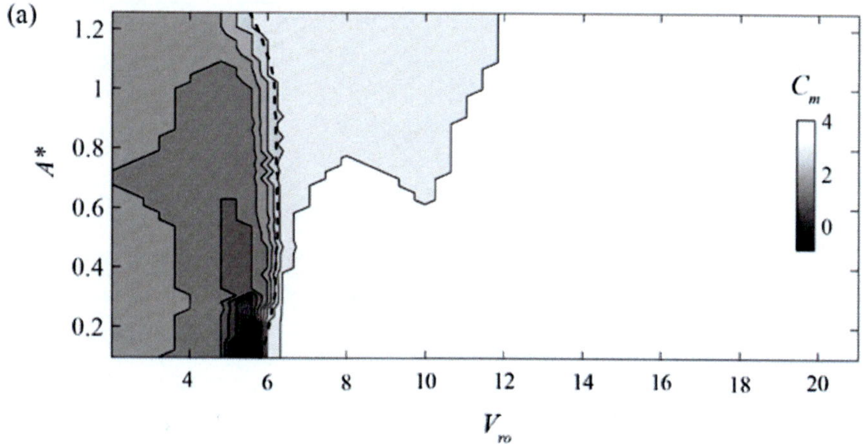

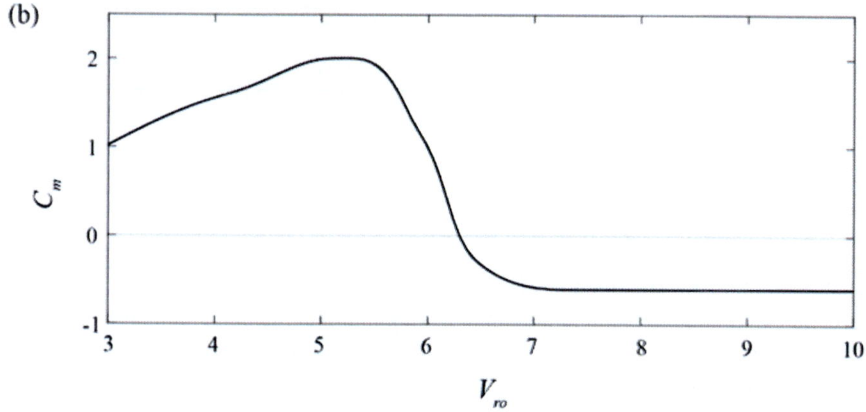

Fig. 4.8 Added mass coefficient, C_m, (**a**) versus the amplitude of imposed oscillations and V_{ro}, and (**b**) versus V_{ro} for a constant amplitude of 0.6. The dashed lines represent zero. (Reconstructed from Gopalkrishnan, 1993)

shedding patterns more complicated than the 2S pattern. The 2S and 2P patterns that are observed here are in the low and high values of V_{ro}, respectively, similar to what we observed in the wake of the self-excited VIV tests discussed in Sect. 6.

9 Modeling VIV

In this section, we discuss how we can model a VIV system by considering one of the existing models for VIV. We will not discuss methods that rely on computational fluid dynamics (CFD) simulations to model VIV: that is solving the flow around the cylinder using CFD methods and then coupling the flow forces with the structure's

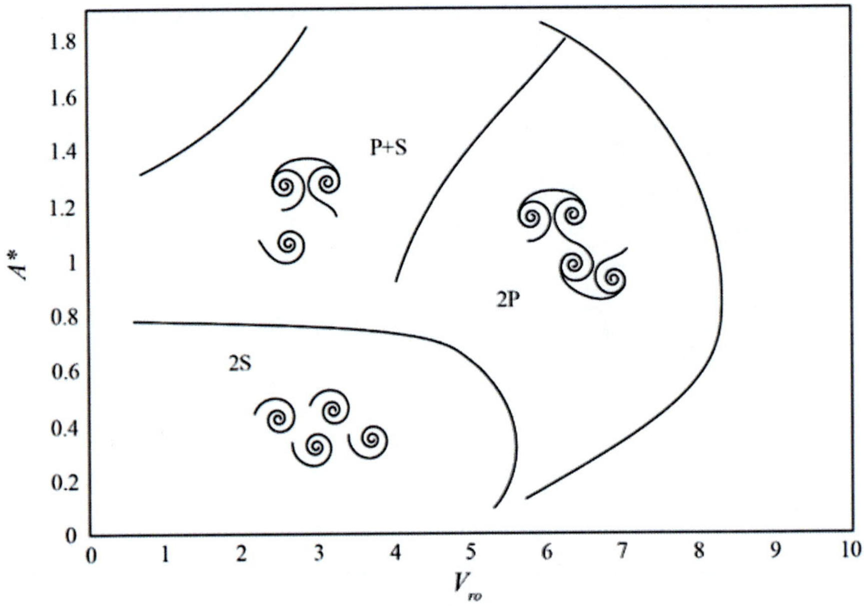

Fig. 4.9 A map of shedding patterns in the wake of a cylinder forced to oscillate with prescribed amplitudes and frequencies. S: Single vortex, P: Pair of vortices. (Reproduced from Williamson & Govardhan, 2004)

response. We should be aware that those methods do exist though. Here, we focus on the models that are called phenomenological models of the problem. Instead of solving the Navier-Stokes equations, we model the flow around the cylinder based on our understanding of VIV. These models are appropriate tools to obtain a better understanding of a phenomenon, instead of predicting the system behavior exactly. The VIV model that we discuss here was first derived by Facchinetti et al. (2004). There are also other models for VIV as reviewed by Facchinetti et al. (2004) and Païdoussis et al. (2010). The symbols are different here from what they have used to be consistent with what we have done here so far.

To start our modeling approach, we consider the original equation of motion for a 1D VIV system, which has the form of Eq. (4.3)

$$\ddot{y} + 2\zeta\omega_n\dot{y} + \omega_n^2 y = \frac{1}{2}\frac{\rho}{m}DU^2 c_y \sin(\omega_s t + \phi). \tag{4.40}$$

or Eq. (4.37):

$$\left(m + C_m\rho\frac{\pi D^2}{4}\right)\ddot{y} + \left(c - \frac{1}{2}\rho D U^2 C_{LV}\right)\dot{y} + ky = 0. \tag{4.41}$$

As we discussed before, the challenge in solving this equation is that C_y and ω_s in Eq. (4.40) or C_m and C_{LV} in Eq. (4.41) depend on the system's response, which is the solution of the equation. All the parameters on the left-hand side of Eq. (4.40) depend on the structure and are known once we define our system. If, somehow, we relate the flow forces on the right-hand side of Eq. (4.40) to the system's response (i.e., $y, \dot{y}, \ddot{y}$), through a relation with known coefficients, then we will be able to solve the governing equation. Of course, this will be based on some assumptions. Here, we should remember the limit cycle nature of the VIV response that we discussed in Sect. 8: Small-amplitude oscillations result in a negative flow-induced damping, and an increase in the amplitude, and large-amplitude oscillations result in a positive flow-induced damping, and a decrease in the amplitude. This limit cycle nature has led to models based on a *wake oscillator* for VIV systems. In these models, the wake of the cylinder is assumed to be a separate vibratory system with its own equation of motion. This new equation is supposed to behave similarly to the wake, but it has a simpler form than the Navier-Stokes equations, which are the true representation of the wake. An equation which can produce a limit cycle solution is normally selected as a wake oscillator. This means that the wake will undergo a limit cycle oscillation, and because it is in contact with the structure, it will force the structure to follow a limit cycle oscillation, which is what we expect to observe in a system undergoing VIV.

First, we make Eq. (4.40) dimensionless using the following dimensionless parameters:

$$f_n^* = \frac{f_n}{f_s}, Y = \frac{y}{D}, T = 2\pi t f_s, \tag{4.42}$$

where $f_n = \omega_n/2\pi$ is the natural frequency of the structure with no fluid around it, f_s is the frequency of vortex shedding, and D is the cylinder's diameter. By using these dimensionless parameters, we can rewrite the cylinder's velocity and acceleration as:

$$\dot{y} = \frac{\partial y}{\partial t} = \frac{\partial (DY)}{\partial \left(\frac{T}{2\pi f_s}\right)} = 2\pi f_s D \frac{\partial Y}{\partial T} = 2\pi f_s D Y', \tag{4.43}$$

and

$$\ddot{y} = \frac{\partial^2 y}{\partial t^2} = \frac{\partial^2 (DY)}{\partial \left(\frac{T}{2\pi f_s}\right)^2} = 4\pi^2 f_s^2 D \frac{\partial^2 Y}{\partial T^2} = 4\pi^2 f_s^2 D Y'', \tag{4.44}$$

where $(\ )'$ denotes derivative with respect to dimensionless time, T.

Plugging Eqs. (4.43) and (4.44) into Eq. (4.40), we find:

$$4\pi^2 f_s^2 Y'' + 2(2\pi)\zeta\omega_n f_s Y' + \omega_n^2 Y = \frac{1}{2}\frac{\rho}{m} c_y U^2 \sin\left(\omega_s \frac{T}{2\pi f_s} + \phi\right). \tag{4.45}$$

Then we use $\omega_n = 2\pi f_n$, $\omega_s = 2\pi f_s$ and $f_s = StU/D$ to come up with the following form:

$$4\pi^2\left(St\frac{U}{D}\right)^2 Y'' + 2(2\pi)\zeta(2\pi f_n)St\frac{U}{D}Y' + (2\pi f_n)^2 Y$$
$$= \frac{1}{2}\frac{\rho}{m}c_y U^2 \sin(T+\phi), \tag{4.46}$$

and then we divide everything by $4\pi^2(St\,U/D)^2$ to find:

$$Y'' + 2\zeta\frac{1}{St}\frac{f_n D}{U}Y' + \frac{f_n^2 D^2}{St^2 U^2}Y = \frac{1}{2}\frac{\rho}{m}c_y\frac{D^2}{4\pi^2(St)^2}\sin(T+\phi). \tag{4.47}$$

The inverse of the reduced velocity, $f_n D/U \equiv 1/V_r$, shows up in this equation, which means that we can write the equation as:

$$Y'' + 2\zeta\frac{1}{St}\frac{1}{V_r}Y' + \frac{1}{St^2}\frac{1}{V_r{}^2}Y = \frac{1}{2}\frac{\rho}{m}c_y\frac{D^2}{4\pi^2(St)^2}\sin(T+\phi). \tag{4.48}$$

If we call

$$\frac{1}{St}\frac{1}{V_r} \equiv \delta, \tag{4.49}$$

then we get the following form for the equation:

$$Y'' + 2\zeta\delta Y' + \delta^2 Y = \frac{1}{2}\frac{\rho}{m}c_y\frac{D^2}{4\pi^2(St)^2}\sin(T+\phi), \tag{4.50}$$

and since $\frac{\pi}{4}\rho D^2/m = 1/m^*$:

$$Y'' + 2\zeta\delta Y' + \delta^2 Y = \frac{1}{2}\frac{1}{m^*}c_y\frac{1}{\pi^3(St)^2}\sin(T+\phi). \tag{4.51}$$

The right-hand side of this equation represents the flow forces that act on the structure, and it is a function of the added mass, m^*, Strouhal number, St, and most importantly, the coefficient of the crossflow forces, c_y. The main challenge in modeling a VIV system is to properly represent this right-hand side, which we do here by using the van der Pol wake oscillator model. We assume that the right-hand side has two parts: one proportional to Y', i.e., the flow-induced damping of the system, and one proportional to an unknown parameter, q, which represents the rest of the flow forces. This gives us the following form for the right-hand side:

$$\frac{1}{2}\frac{1}{m^*}c_y\frac{1}{\pi^3(St)^2}\sin(T+\phi)\equiv-\frac{\gamma}{m_1^*}Y'+\frac{M}{m_1^*}q, \tag{4.52}$$

in which, γ and M are constants (note: M is not a mass) and m_1^* has a form similar to what we called the mass ratio before (This is not exactly the way we defined the mass ratio before (Eq. (4.11)), and that is why we call it m_1^* instead of m^*.):

$$m_1^*=\frac{m+m_f}{\rho D^2}, \tag{4.53}$$

in which m_f is the added mass, $m_f = C_m\rho A$, where A is the cross-sectional area of the cylinder. In this model, a value of $C_m = 1$ is used (remember this is equal to the added mass coefficient around a cylinder moving in otherwise still fluid found using a potential flow assumption).

In the first term of Eq. (4.52), γ controls the magnitude of the flow-induced damping and is related to the drag coefficient as:

$$\gamma=\frac{C_D}{4\pi St}, \tag{4.54}$$

and with the assumption that $C_D = 2$ and $St = 0.2$, we find a value of $\gamma = 0.8$ in this model.

In the second term of Eq. (4.52), M controls the magnitude of the fluctuating flow force acting on the structure and is related to the magnitude of the lift force as:

$$M=\frac{C_{L_0}}{2}\frac{1}{8\pi^2 St^2}, \tag{4.55}$$

where C_{L_0} is the fluctuating lift coefficient for a fixed cylinder placed in flow. For a constant $C_{L_0} = 0.3$ and $St = 0.2$, we find a value of $M = 0.05$ in this model.

The relations for γ and M in Eqs. (4.54) and (4.55) are found based on curve fitting of some experimental data, and the values for C_D, C_{L_0}, and St are also based on some experimental data. Clearly, the results of this model depend heavily on these relations and these parameters.

With the form of Eq. (4.52) for the right-hand side of Eq. (4.51), the equation of motion becomes:

$$Y''+\left(2\zeta\delta+\frac{\gamma}{m_1^*}\right)Y'+\delta^2 Y=\frac{M}{m_1^*}q. \tag{4.56}$$

To find q, we use a wake oscillator model based on a van der Pol equation. This is a nonlinear equation and with a certain set of parameters it results in limit cycle oscillations. Since this equation is supposed to represent the wake, it should be affected by the cylinder's motion. To show this influence, we need to have some sort of a structural dependency in the van der Pol equation, which we do by using an external force on the right-hand side of the van der Pol equation, which is a function of the cylinder acceleration, Y'':

$$q'' + \varepsilon(q^2 - 1)q' + q = AY''. \tag{4.57}$$

What we see on the left-hand side of Eq. (4.57) is the van der Pol equation, and the right-hand side is an external force. Note that we are using the dimensionless form of the acceleration on the right-hand side, and the dimensionless time in all time derivatives. The two constants, A and ε, are obtained based on fitting some curves to the experimental data, and are found to be $A = 12$ and $\varepsilon = 0.3$.[1] The source of nonlinearity in Eq. (4.57) is the damping term: $\varepsilon(q^2 - 1)q'$. Assume ε is a positive number. For a small value of q, q^2 is smaller than 1, resulting in a negative value for $q^2 - 1$, and therefore a negative damping (i.e., increasing q). For a large value of q, q^2 is larger than 1, resulting in a positive value for $q^2 - 1$, and therefore a positive damping (i.e., decreasing q). This nonlinear term, therefore, gives us the limit cycle behavior that we expect from the wake.

In summary, for our wake oscillator model, we have the following set of equations to solve and find the VIV response of a flexibly mounted rigid cylinder:

$$Y'' + \left(2\zeta\delta + \frac{\gamma}{m_1^*}\right)Y' + \delta^2 Y = \frac{M}{m_1^*}q, \tag{4.58}$$

$$q'' + \varepsilon(q^2 - 1)q' + q = AY''. \tag{4.59}$$

Once we know our system (i.e., its mass per unit length, diameter, natural frequency, etc.) then all the parameters in these two equations are known, and the only unknowns are Y and q. Here, we use this set of equations and solve them for a system with $\zeta = 0.002$ and $m_1^* = (m + \rho C_m \pi D^2/4)/(\rho D^2) = (0.59 + 1000 \times \pi \times 0.0127^2/4)/(1000 \times 0.0127^2) = 4.4$. We also get $\delta = 1/(StV_r) = 1/(0.2V_r) = 5/V_r$, where we vary V_r over a range, and we know that for this wake oscillator model we always have $\gamma = 0.8$, $M = 0.05$, $A = 12$, and $\varepsilon = 0.3$. We vary V_r from a small value to a desired value, in this case up to $V_r = 13$. For each reduced velocity, we solve Eqs. (4.58) and (4.59) and we find Y and q. Figure 4.10 shows the time histories of Y and q for $V_r = 6$, where q represents the fluctuating flow force that acts on the cylinder and Y is the cylinder's response. As expected, they are both periodic functions and there is a phase difference between them. The phase difference shows how it takes some time for the cylinder to respond to the flow forces. The frequency of oscillations for both time histories are the same, because we are in the lock-in range and we expect the fluctuating fluid forces and the cylinder's oscillations to be synchronized.

[1] Here we avoid the details of how exactly this is done, but you can read more about it in Facchinetti et al. (2004).

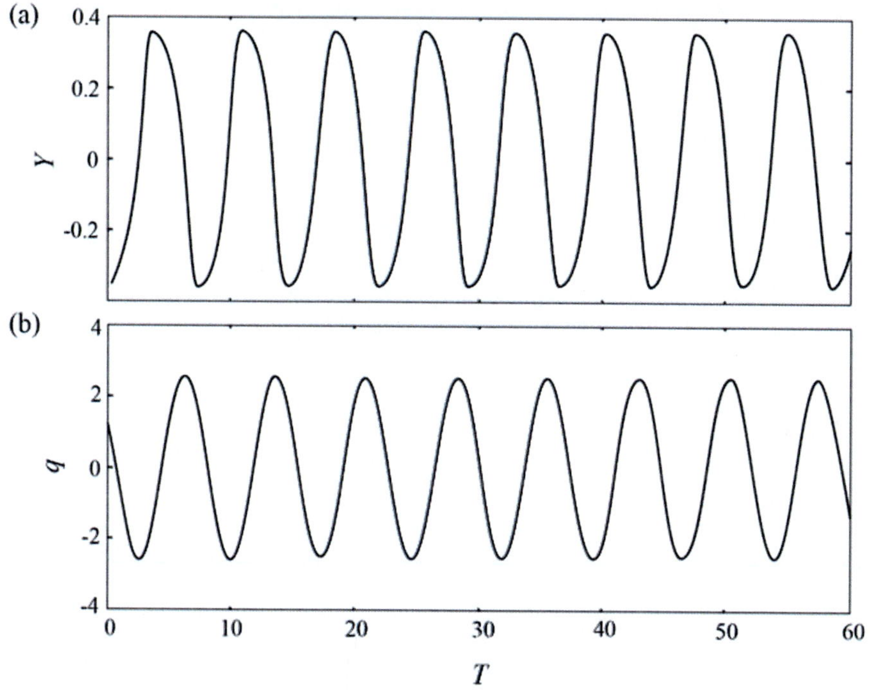

Fig. 4.10 Time histories of Y and q for a reduced velocity of $V_r = 6$

If we solve Eqs. (4.58) and (4.59) for every reduced velocity, then we can plot the amplitude of response versus the reduced velocity. This is shown in Fig. 4.11 based on the wake oscillator model. This plot clearly shows the lock-in range. The mass-damping parameter in this case ($m^* \zeta = 0.0073$) is smaller than the mass-damping parameter of the system in Fig. 4.2, the width of the lock-in range and the amplitude of response in Fig. 4.11 are also smaller than the experimental results of Fig. 4.2. In general, quantitatively, we should not rely on results from a van der Pol wake oscillator model very much, and we should use this model for qualitative understanding of VIV. In particular, the wake oscillator models are useful tools to study the relative influence of different parameters on the VIV response.

As an example, we can use the wake oscillator model to study how the mass-damping parameter changes the magnitude and the width of the lock-in range. Conducting this study experimentally or using CFD-based methods is very time-consuming, but the van der Pol wake oscillator model is capable of doing it very quickly. Figure 4.12 shows the amplitude versus the reduced velocity for different values of the mass-damping parameters. We have changed these by varying the mass ratio of the system. It is clear that as we increase the mass-damping parameter, the maximum amplitude of the response decreases, and the lock-in range becomes

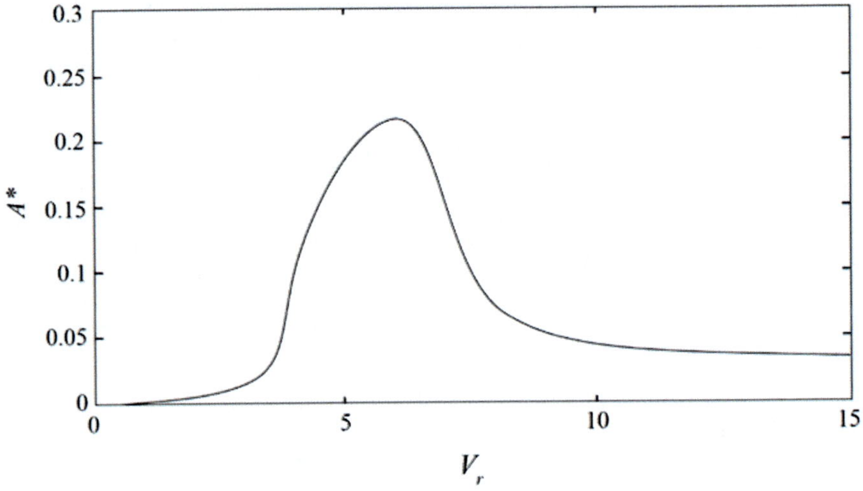

Fig. 4.11 The dimensionless amplitude versus the reduced velocity found using the Van der Pol wake-oscillator model

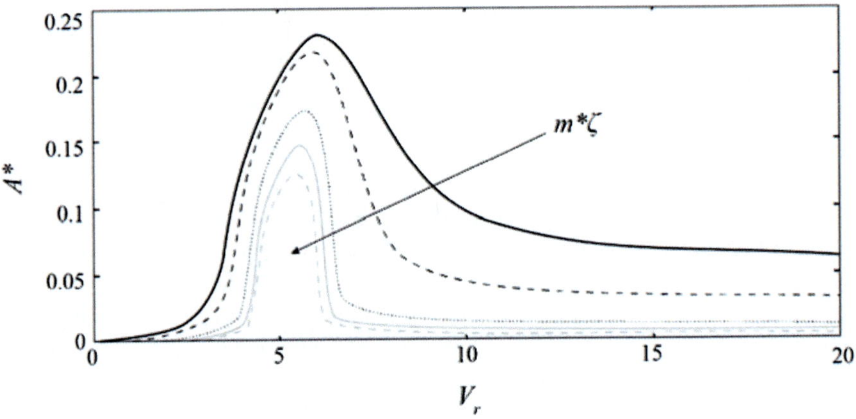

Fig. 4.12 The amplitude versus the reduced velocity for different values of the mass-damping parameter, $m^*\zeta$

narrower. We had seen the decrease in the maximum amplitude of the response based on the experimental results in Fig. 4.4 as well. While those results were based on several hours of experimental effort, the results of Fig. 4.12 were obtained on a PC in minutes.

10 VIV in the Inline Direction

In this section, we consider a 1 DOF mass-spring-damper system again, but this time we place it in flow such that it is allowed to oscillate in the direction of flow (inline – IL), instead of CF (Fig. 4.13). The equation of motion for this system is:

$$m\ddot{x} + c_x\dot{x} + k_x x = F_x, \tag{4.60}$$

where the external force is the fluctuating force that acts on the structure from the fluid in the direction of flow:

$$F_x = \frac{1}{2}\rho D U^2 c_x \sin\left(2\omega_s t + \phi_{IL}\right). \tag{4.61}$$

The frequency of this external force is $2\omega_s$, same as the frequency of fluctuating drag that we discussed in Chap. 2. Then the equation of motion (Eq. (4.60)) can be written in terms of the natural frequency and the damping ratio as:

$$\ddot{x} + 2\zeta\omega_{n\_IL}\dot{x} + \omega_{n\_IL}^2 x = \frac{1}{2}\frac{\rho}{m}c_x D U^2 \sin\left(2\omega_s t + \phi_{IL}\right), \tag{4.62}$$

where $\omega_{n\_IL} = \sqrt{k_x/m}$ is the natural frequency of the system. Note that we use ω_{n_IL} to refer to the natural frequency of a system that oscillates in the IL direction, and ω_n to refer to the natural frequency of a system that oscillates in the CF direction.

Figure 4.14 shows the amplitude and frequency of a typical response of this system. As we increase the flow velocity from zero, the shedding frequency increases linearly with flow velocity following the Strouhal law. Since the external force that acts on the structure has a frequency twice the shedding frequency, we expect to observe synchronization between the external force and the natural frequency of the system when $2\omega_s = \omega_{n\_IL}$, or $\omega_s = \frac{1}{2}\omega_{n\_IL}$. In the plot of frequency versus reduced velocity (Fig. 4.14b), this point corresponds to the intersection of a horizontal line that represents the natural frequency and a line that corresponds to $2St$ (very close to $V_r = 2.5$). At this point, oscillations are observed in the system for a range of reduced velocities. This is shown in the amplitude plot of Fig. 4.14a for the

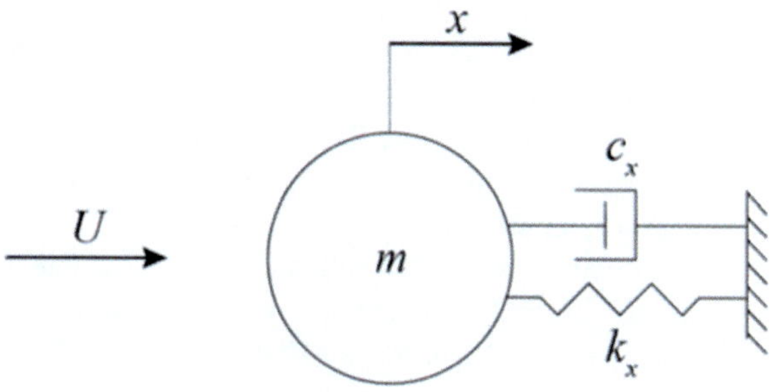

Fig. 4.13 A 1 DOF mass-spring-damper placed in flow and allowed to oscillate in the direction of flow (inline – IL)

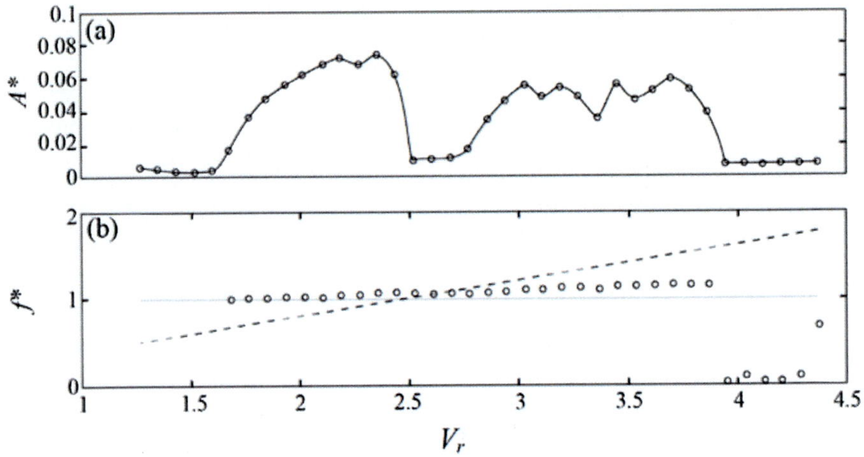

Fig. 4.14 Dimensionless (**a**) amplitude, A^*, and (**b**) frequency, f^*, of oscillations versus V_r for a cylinder undergoing IL VIV. (Data from Gurian et al., 2019)

reduced velocity range of $V_r \sim 2.8$–3.7. The amplitude of oscillations is around 0.05D—much smaller than what we observed in the CF VIV response. With a Strouhal number of $St = 0.2$, in the CF VIV, we observed an onset of lock-in at approximately $V_r = 1/St = 5$. For the IL VIV, we expect the oscillations to lock in with the shedding frequency at $2St$, which would correspond to $V_r = 1/2St = 2.5$, and this is what we have observed as we just discussed. However, in the amplitude plot of Fig. 4.14a, we observe another range of nonzero-amplitude oscillations in the reduced velocity range of $V_r = 1.8$–2.4. According to what we discussed above, we do not expect any oscillations at this point. At $V_r = 1.8$, the shedding frequency is not equal to the system's natural frequency, and the shedding should follow the Strouhal law and keep increasing linearly. This would have been the case, if we had no disturbance in the system. But if we disturb the cylinder by giving it an initial displacement, it moves in the IL direction and sheds a pair of symmetric vortices. The shed vortices then exert a force on the cylinder, and the cylinder keeps moving and sheds another pair of symmetric vortices in the wake. This scenario is repeated and the cylinder's oscillations become synchronized with the shedding of symmetric vortices.

Figure 4.15 shows the wake of the cylinder in both nonzero-amplitude regions. In the first region, the symmetric shedding that we just discussed is observed. The two vortices have very similar sizes and are shed at the same time from both sides of the cylinder in each cycle of oscillations. In the second region, an asymmetric shedding is observed, where one vortex is shed from one side during one cycle of oscillations and another vortex is shed from the other side during the next cycle of oscillations.

In summary, for a 1 DOF cylinder free to oscillate in the IL direction, we observe two regions of oscillations: in the first region, oscillations start only if the structure is triggered. In this region, which starts at a reduced velocity of $V_r \sim 1.8$, symmetric

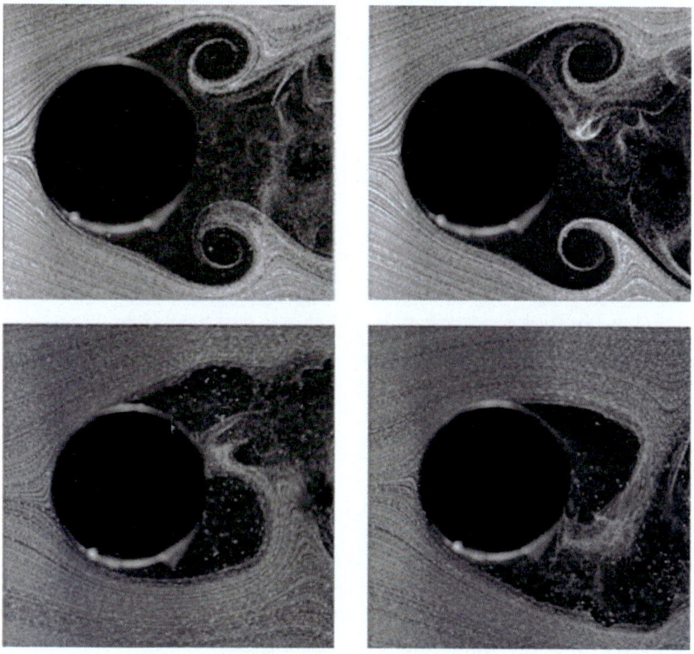

Fig. 4.15 (top) Symmetric and (bottom) asymmetric shedding observed in the wake of a cylinder undergoing IL VIV in the first and the second regions of oscillations. (From Gurian et al., 2019)

vortices are observed in the wake. The second region of oscillations starts when the shedding frequency synchronizes with the system's natural frequency. Since the fluctuating force frequency in the IL direction is twice the shedding frequency, the onset of this second region of oscillations is at a reduced velocity of $V_r \sim 2.5$. In this region, asymmetric shedding is observed in the wake. In both regions, the amplitude of oscillations stays around 0.1D—much smaller than the amplitude of oscillations in CF VIV.

11 Two-Degree-of-Freedom VIV

After discussing VIV of systems with one degree of freedom (crossflow or inline), in this section, we consider VIV of systems with two degrees of freedom. To do so, we consider a mass-spring-damper system with two degrees of freedom placed in fluid flow, such that it can oscillate both in the CF and IL directions (Fig. 4.16). This system has one mass but two sets of springs and dampers, which means that the mass of the structure is constant in both directions, but the stiffness and damping could be different. Jauvtis and Williamson (2004) and Dahl et al. (2006, 2007, 2010), among others, have discussed 2D VIV with some details.

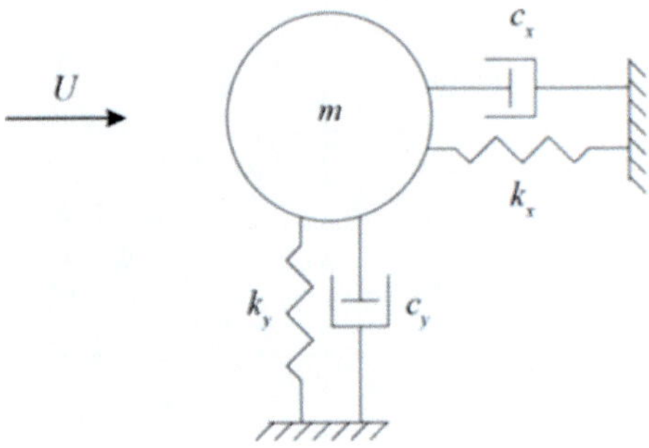

Fig. 4.16 A two-degree-of-freedom mass-spring-damper placed in flow

The equations of motion for this system are two second-order ODEs:

$$m\ddot{x} + c_x\dot{x} + k_x x = F_x(x, \dot{x}, \ddot{x}, y, \dot{y}, \ddot{y}), \tag{4.63}$$

$$m\ddot{y} + c_y\dot{y} + k_y y = F_y(x, \dot{x}, \ddot{x}, y, \dot{y}, \ddot{y}), \tag{4.64}$$

in which c_x and c_y are the damping coefficients in the inline and crossflow directions, k_x and k_y are the stiffness values in the inline and crossflow directions, and F_x and F_y are the flow forces in the inline and crossflow directions, respectively. The major difference compared with the 1 DOF case is that F_x and F_y depend on both the x and the y responses. Therefore, the two equations are coupled, and we cannot solve one by itself and ignore the other one: we must solve the two of them at the same time. We can write the flow forces that appear on the right-hand side of Eqs. (4.63) and (4.64) as

$$F_x = \frac{1}{2}\rho D U^2 c_{IL} \sin(2\omega_s t + \phi_x), \tag{4.65}$$

and

$$F_y = \frac{1}{2}\rho D U^2 c_{CF} \sin\left(\omega_s t + \phi_y\right), \tag{4.66}$$

where c_{IL} and c_{CF} are the force coefficients in the IL and CF directions, respectively, and ϕ_x and ϕ_y are phase differences between the forces and the displacements in their corresponding directions. Note that the forcing frequency in the CF direction is equal to the shedding frequency, ω_s, and in the IL direction is equal to twice the shedding frequency, $2\omega_s$. This is because the fluctuating drag frequency is twice the shedding frequency as we discussed in Chap. 2.

Then using Eqs. (4.65) and (4.66), we can write Eqs. (4.63) and (4.64) as:

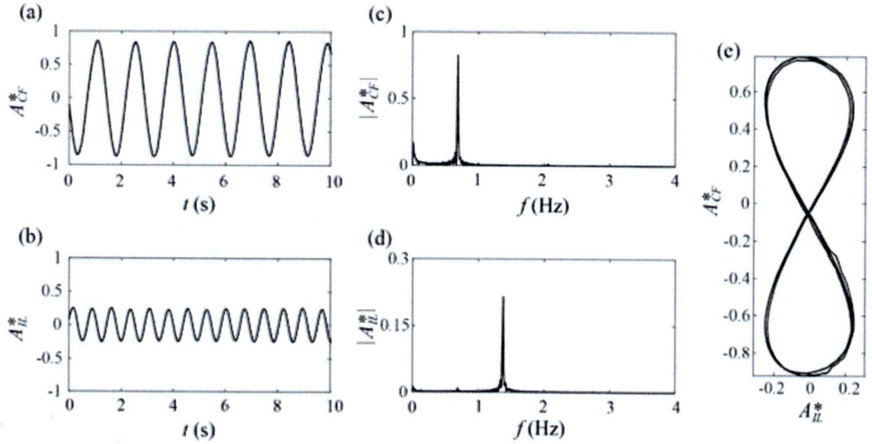

Fig. 4.17 Sample VIV response of a 2 DOF cylinder placed in flow: Displacements and frequency contents in the (**a**, **b**) CF and (**c**, **d**) IL directions, together with (**e**) the response trajectory (Carlson et al., 2021)

$$\ddot{x} + 2\zeta_{IL}\omega_{n\_IL}\dot{x} + \omega_{n\_IL}^2 x = \frac{1}{2}\frac{\rho}{m}DU^2 c_{IL} \sin\left(2\omega_s t + \phi_x\right), \tag{4.67}$$

and

$$\ddot{y} + 2\zeta_{CF}\omega_{n\_CF}\dot{y} + \omega_{n\_CF}^2 y = \frac{1}{2}\frac{\rho}{m}DU^2 c_{CF} \sin\left(\omega_s t + \phi_y\right), \tag{4.68}$$

in which ζ_{IL} and ζ_{CF} are the structural damping ratios, and ω_{n_IL} and ω_{n_CF} are the system natural frequencies in vacuum in the IL and CF directions, respectively.

If we choose the spring constants in the two directions in a way that the structural natural frequency in the IL direction becomes twice the structural natural frequency in the CF direction, i.e., $\omega_{n\_IL} = 2\omega_{n\_CF}$, then once the CF forcing frequency equals the CF natural frequency, the IL forcing frequency will equal the IL natural frequency as well, which results in large amplitude oscillations in both directions at the same time. This is sometimes referred to as dual resonance (Dahl et al., 2010). Clearly, the ratio between the natural frequencies in the IL and CF directions does not have to be 2:1. However, here, we consider a case with a 2:1 ratio. A sample of such oscillations in both the CF and IL directions is shown in Fig. 4.17. The frequency of oscillations in the IL direction is twice that in the CF direction. This is obvious both in the time histories and FFT plots of Fig. 4.17. The amplitude of the CF oscillations is larger than that of the IL oscillations, but the IL amplitude is not negligible—it is larger than the amplitude observed in the pure IL VIV. If we plot the CF displacement versus the IL displacement (Fig. 4.17e), we observe a figure-eight trajectory for the cylinder's motion, as expected, since the frequency ratio of the IL and CF oscillations is 2:1. When we consider the trajectory plot, it is also important

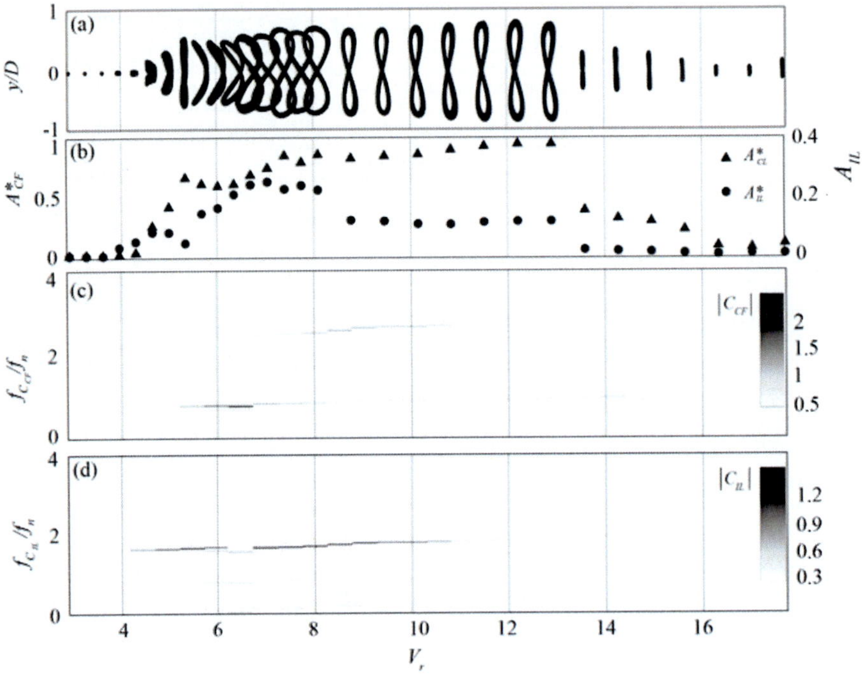

Fig. 4.18 The response of a 2D VIV system: (**a**) the trajectories, (**b**) the CF and IL amplitudes, (**c**) the normalized CF, and (**d**) IL force frequencies versus the reduced velocity (Carlson et al., 2021)

to know the direction of flow. In the plot of Fig. 4.17e, the flow is from left to right and the cylinder moves opposite the flow direction at the two extreme ends of its CF motion, i.e., at the highest and the lowest points of the trajectory. This direction of the trajectory is sometimes referred to as a counterclockwise (CCW) direction, since the motion of the upper lobe of the figure-eight is in the CCW direction. The CCW orbital motion is extensively observed in the VIV response of bluff bodies with two degrees of freedom.

Figure 4.18 shows the amplitude plots and the force-frequency contents of the response in both the CF and IL directions versus the reduced velocity for a sample 2D VIV case, with a 2:1 ratio of the IL and CF natural frequencies. The figure-eight trajectory is observed for the entire lock-in range (Fig. 4.18a), and for this entire range, the cylinder moves upstream at the two extremes of its trajectory, i.e., a CCW figure-eight trajectory, same as what we saw in Fig. 4.17. Oscillations both in the IL and CF directions start at $V_r \sim 3.6$. The amplitudes of these oscillations are very small, but their existence can be verified by observing the wake at this reduced velocity (Fig. 4.19a). During each cycle of oscillations two vortices are shed simultaneously from the two sides of the cylinder. For a cylinder fixed in flow, a von Kármán vortex street is expected, in which asymmetric vortex shedding is observed. If the cylinder oscillates mainly in the CF direction, then a 2S (two single

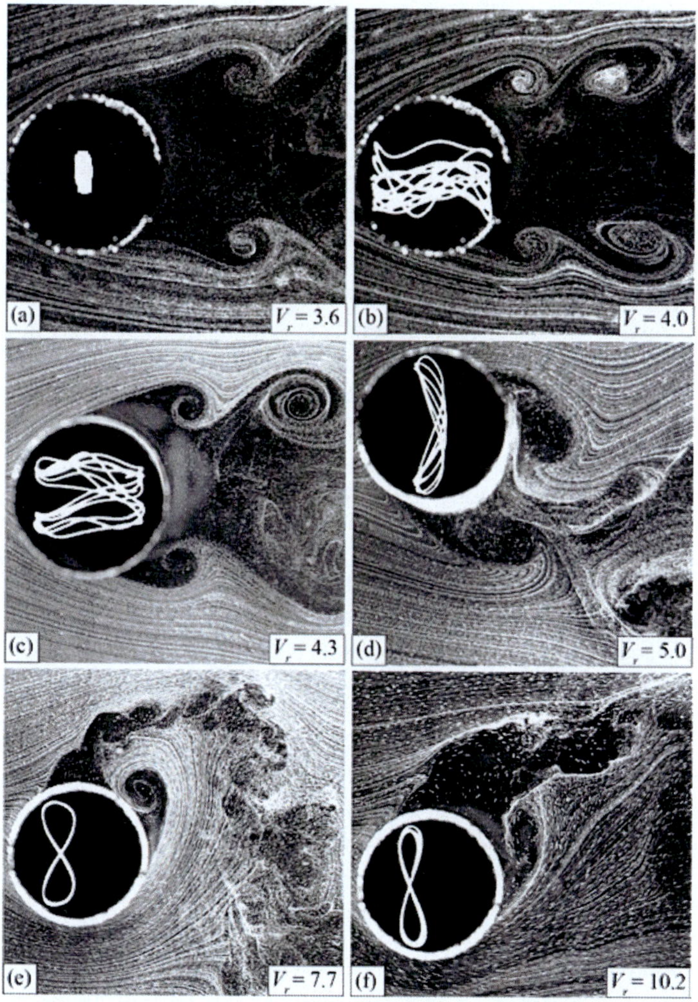

Fig. 4.19 The wake of a cylinder undergoing 2D VIV at six sample reduced velocities (Carlson et al., 2021)

vortices shed per cycle of oscillations, one after the other) is expected. However, here two symmetric vortices are shed simultaneously from the two sides of the cylinder. This type of symmetric shedding is also observed in IL VIV, when oscillations exist only in the IL direction, as we discussed in Sect. 10. The presence of these symmetric vortices here confirms that oscillations in the inline direction exist at this reduced velocity. In a 1 DOF CF system, no oscillation is expected to start before $V_r \sim 5$. Note that the reduced velocity of $V_r \sim 3.6$ (for which oscillations start in the plots of Fig. 4.18) is defined based on the CF frequency. The corresponding reduced velocity defined based on the IL frequency, which is normally used in studies of IL VIV, would be half of that of the CF-frequency-based V_r, i.e., V_{r_IL} ~1.8. This reduced velocity is within the first range of oscillations for a cylinder in pure IL VIV. According to the IL VIV studies, the first range of oscillations is observed for V_{r_IL} ~ 1.7–2.5, which corresponds to V_r ~ 3.4-5. In the plots of Fig. 4.18, we do observe a mainly IL oscillations for this range of reduced velocities, which results in the shedding of symmetric vortices in the wake (Fig. 4.19).

Another example of these oscillations ($V_r = 4$), with a larger amplitude, is shown in Fig. 4.19b. We should note that oscillations in the CF are not expected within this range of reduced velocities in the pure CF VIV cases. The synchronization in the IL direction, however, also causes some oscillations in the CF direction, with a smaller magnitude. At higher reduced velocities (V_r = 4.3 – Fig. 4.19c), while the IL oscillations are still the main observed oscillations, CF oscillations become more significant, but their magnitude is still approximately half that of the IL oscillations, and the shedding remains symmetric. At this reduced velocity, a figure-eight type trajectory starts to appear in the response of the system. This is the emergence of the figure-eight trajectories that are observed extensively in the VIV response of a 2 DOF cylinder, as a result of the 2:1 frequency ratio between the IL and CF oscillations. The figure-eight at this reduced velocity is a "wide" figure-eight with the IL oscillations dominating the response.

At V_r ~ =5, the figure-eight becomes narrower, resembling more the type of figure-eight that is expected in a typical 2D VIV response: with a larger CF amplitude than the IL amplitude. The wake at this reduced velocity shows features of both the pure IL and the pure CF oscillations: At some instances of oscillations, symmetric sheddings in the IL direction appear, but they are influenced by the asymmetric sheddings of larger vortices as a result of the large-amplitude CF oscillations. The result is shedding of asymmetric pairs of vortices in the wake (Fig. 4.19d).

The amplitude of the CF oscillations increases rapidly from $V_r = 4.5$ to 5.5, while the amplitude of the IL oscillations also increases slightly (Fig. 4.18b). At V_r ~ 5.5, which is approximately the reduced velocity at which the lock-in is expected to start for a purely CF VIV case, a local maximum is observed in the CF amplitude, followed by a slight decrease in the magnitude, and then another increase, leading to a range of reduced velocities (V_r = 8–13) for which the CF amplitude remains constant and close to $A^* = 0.9$. The IL amplitude follows a bell-shaped curve for a range of V_r = 5–9, corresponding to the range where the CF amplitude drops from its first local maximum and then increases again. The range of V_r = 5.5–8 corresponds to the V_{r_IL} = 2.75–4. Curiously, this is the range of reduced velocities for which in a purely IL VIV case, a second range of nonzero amplitudes is observed. In this range, for a pure IL VIV case, oscillations of up to 0.05D are observed, together with asymmetric shedding in the wake. The 2:1 ratio between the IL and CF frequencies makes this range of reduced velocities correspond also to the beginning of the expected range of lock-in for a purely CF case. Therefore, in the 2 DOF case, for the range that the pure IL VIV would also be expected, the IL amplitude does follow its bell-shaped case and causes fluctuations in the amplitude of the CF oscillations. This is the manifestation of the dual resonance as discussed by Dahl et al. (2007). In this range of flow velocities, at the beginning of each half cycle of oscillations, a shear layer is separated from one side of the cylinder. The shear layer bends quickly and forms an attached vortex, which travels with the cylinder to the end of that half-cycle, and is then shed (Fig. 4.19e).

After the range of expected IL oscillations in the pure IL case ends ($V_r = 8$), oscillations in the CF are still expected, because $V_r = 8$ is still in the middle of the expected lock-in range for a pure CF case, and IL oscillations are observed here as a result of the CF synchronization. In a pure IL case, no oscillations would have been observed in this range. The IL oscillations in this range of reduced velocities are with a relatively constant magnitude. The trajectory is still a figure-eight. The wake (Fig. 4.19f) is very similar to the wake in the previous range.

At $V_r \sim 13.5$, the IL oscillations end, which results in a significant drop in the CF amplitude. The amplitude of CF oscillations continues to decrease with increasing reduced velocity, until these oscillations also end at $V_r = 16$.

During the range of reduced velocities where both pure IL and pure CF VIV are expected ($V_r = 5.5$–8), the trajectory transforms from a figure-eight that is significantly bent to the left (toward the incoming flow) to a straight figure-eight. When the IL oscillations stay constant (for $V_r = 8$–13) the figure-eight orbits stay straight, until they turn into straight lines when the IL oscillations end at $V_r \sim 13.5$.

These different ranges of response are also clear in the frequency contents of the CF and IL forces (Fig. 4.18c, d). For the range of reduced velocities where oscillations are mainly in the IL direction ($3.6 < V_r < 5$), the main peak frequency is observed in the IL direction and at a normalized frequency of around 2. This is then followed by a range where lock-in is observed in both the CF and IL directions ($5 < V_r < 8$), where peak frequencies in the CF direction are observed both for the first and the third harmonics, and in the IL direction at two times the fundamental frequency. The maximum contribution of the third harmonic is observed in this range of reduced velocities, where dual resonance (simultaneous lock-in in the CF and IL directions) occurs. When the dual resonance ends, and in the range where the response is ruled by the CF oscillations ($8 < V_r < 13$), while the IL oscillations still exist, mainly the first harmonic in the CF and the second harmonic in the IL directions are observed—the third harmonic component in the CF direction is less significant for this range, although it does exist. The third harmonic component disappears when the IL oscillations become very small in the range of $13 < V_r < 16$.

12 Bluff Bodies with Noncircular Cross Sections

Although we focused on a bluff body with a circular cross-section, VIV can be observed for bluff bodies with other cross sections as well. As soon as the cross section of the body changes from circular, however, the mean value of the lift force will not necessary be zero. In some cases, this nonzero mean lift force does not change the response of the system dramatically, and we still observe VIV, a lock-in range, and frequency synchronization. In some other cases, however, the change of the mean lift force to a nonzero value in a body with a noncircular cross section might influence the type of the structural response as well, and instead of VIV, we might observe other types of flow-induced oscillations. In the following chapter, we will discuss these cases, where oscillations other than VIV can occur.

Chapter 5
Galloping and Flutter

1 Introduction

In this chapter, we will consider flow-induced oscillations of structures with noncircular cross sections. For these structures, their orientation with respect to the incoming flow influences the mean flow forces that act on the structure, and as opposed to a structure with a circular cross section, the mean flow force in the direction perpendicular to the flow direction (i.e., the mean lift force) is not necessarily zero. Therefore, the nonzero mean flow forces could cause oscillations, if the structure is flexible or flexibly mounted. In these cases, although vortex shedding can be observed, they are due to the fact that a bluff body is placed in flow, and they are not the cause of oscillations. Besides a bluff body with noncircular cross section, a streamlined body placed in flow can experience nonzero mean lift. The most well-known example of a streamlined body in flow with nonzero mean lift force is an asymmetric airfoil, or any airfoil (symmetric or asymmetric) placed at an angle with respect to the incoming flow. In this chapter, we will see how these nonzero mean values of lift and drag can cause oscillations in flexibly mounted structures. These oscillations can be galloping or flutter. It ought to be mentioned here that flexibly mounted structures with asymmetric cross section could also undergo VIV, similar to what we discussed in Chap. 4, but they do have the potential of undergoing galloping and flutter as well, due to the fact that they experience nonzero mean lift, as we will see in this chapter.

Y. Modarres-Sadeghi, *Introduction to Fluid-Structure Interactions*,
https://doi.org/10.1007/978-3-030-85884-1_5

2 The Mean Flow Forces Acting on a 1 DOF Bluff Body with a Circular Cross Section

Before we discuss a bluff body with a noncircular cross section, we consider a body with a circular cross section (a cylinder) to understand how the mean flow forces do not influence the structure's dynamic response.

When a cylinder is placed in flow with a velocity of U and with no motion for the structure (Fig. 5.1), the only mean flow force that acts on the structure is the drag force, $F_{D,mean}$, which is in the same direction as the direction of the relative velocity. There is no mean lift force, because of the symmetry of the structure, $F_{L,mean} = 0$.

If the cylinder is allowed to move in the vertical direction only (a 1 DOF system), then the relative flow velocity observed by the cylinder will make an angle with respect to the incoming flow direction. The drag force that acts on the cylinder will still be in the direction of the relative flow velocity. In Fig. 5.2, we assume that the cylinder moves downward, and we also assume that y and $\dot{y}$ are positive downward to be consistent with the existing literature.[1] In this case, the component of drag force in the vertical direction, F_y, is upward, while the cylinder's velocity is downward. This means that the direction of the flow force opposes the direction of the cylinder's velocity, i.e., F_y always opposes $\dot{y}$. Note that to observe any flow-induced instability, power should go from the flow to the structure, which means that the flow force and the structure's velocity should be in the same direction. In the present case, this power is always negative, which means that the mean flow force cannot excite the structure.

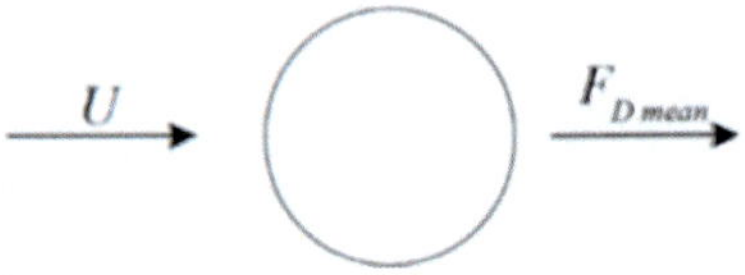

Fig. 5.1 A fixed cylinder in flow

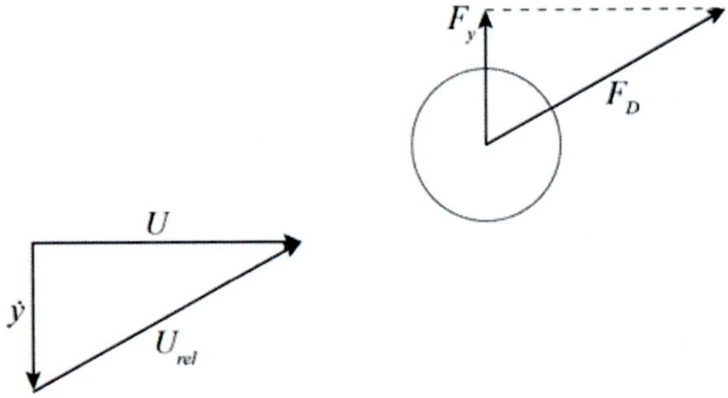

Fig. 5.2 The direction of relative flow velocity observed by the cylinder, U_{rel}, with respect to the incoming flow, together with the direction of the drag force and the force in the y-direction, for a cylinder allowed to move in the vertical, y, direction

[1] We are used to considering + upward, but here we assume + downward. This, of course, does not change the physics, but makes it easier to relate what we find here to the existing studies.

3 1 DOF Oscillations of a Bluff Body with a Noncircular Cross Section

If a bluff body with a noncircular cross section (a prism) is placed in flow, the mean value of the lift force is not necessarily zero, because the cross section is not necessarily symmetric with respect to the incoming flow, and therefore the mean values of both lift and drag forces act on the structure. Figure 5.3 shows an example of such a case. The mean drag and lift forces are in the direction of the relative velocity, U_{rel}, and perpendicular to that, respectively. In this figure, F_y shows the positive direction of the vertical forces. In the case shown in Fig. 5.3, the total force that acts on the structure in the vertical direction, $F_{Total,y}$, is upward and $\dot{y}$ is downward. This means that the flow forces damp the oscillations. However, we can imagine another case in which the mean value of lift acts in the downward direction with a magnitude that can result in an $F_{Total,y}$ in the same direction as $\dot{y}$, as shown in Fig. 5.4. In this case, the flow inputs power into the structure and therefore large-amplitude oscillations can be observed. These large-amplitude oscillations, which are due to the mean values of flow forces acting on the structure, are called galloping.

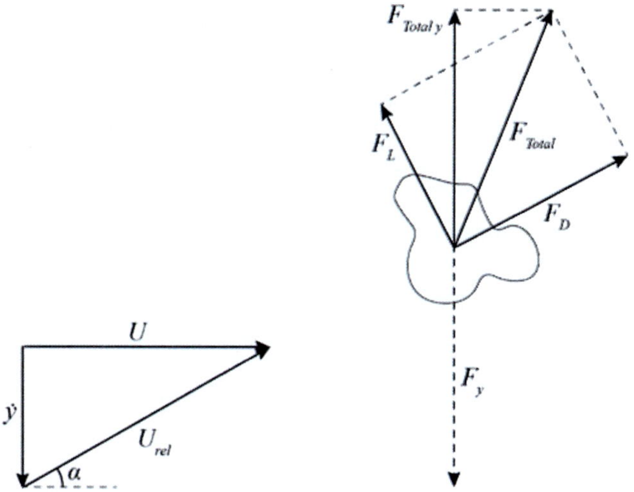

Fig. 5.3 The direction of relative flow velocity observed by the structure, U_{rel}, with respect to the incoming flow, together with the directions of the lift and drag forces and the resultant force in the y-direction, for a structure with a noncircular cross section

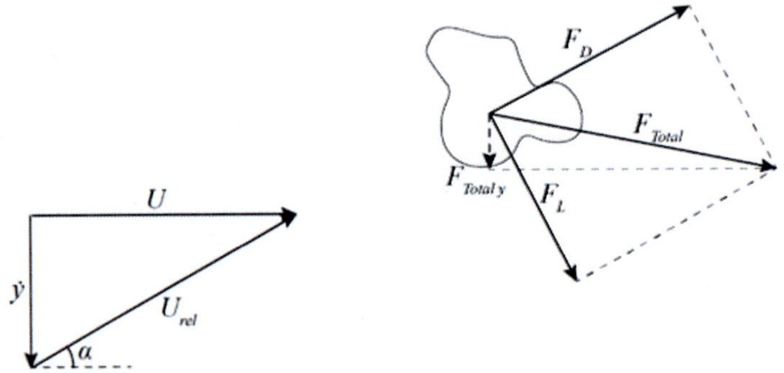

Fig. 5.4 An instant when $F_{Total,y}$ is in the same direction as $\dot{y}$

3.1 *The Equation of Motion for 1 DOF Galloping*

Here, we introduce the equation of motion that describes the motion of the system of Fig. 5.3 to understand how galloping occurs. First, we call the angle between the relative flow velocity seen by the structure and the incoming flow velocity, the angle of attack, α:

$$\alpha = \tan^{-1}\frac{\dot{y}}{U_{rel}}. \tag{5.1}$$

Since in our system the structure can move in the y-direction only (a 1 DOF system), its equation of motion is:

$$m\ddot{y} + 2\zeta m\omega_n\dot{y} + ky = F_y, \tag{5.2}$$

in which F_y represents the flow forces in the y-direction, i.e., the direction of motion. We should remember that y and $\dot{y}$ are positive downward. Also, we should note that the direction of motion is not necessarily the same as the directions of lift or drag forces. The lift and drag forces are perpendicular to the *relative* flow velocity, U_{rel}, but F_y is always in the vertical direction (perpendicular to U), because the degree of freedom is in the vertical direction. We can write F_y in the following form:

$$F_y = \frac{1}{2}\rho h U^2 C_y, \tag{5.3}$$

where h is the height of the prism (perpendicular to the direction of flow) and C_y is the coefficient of flow forces in the vertical direction. This form is similar to what we normally use for lift and drag forces. Since F_y is perpendicular to U, we have used U (not U_{rel}) in the right-hand side of Eq. (5.3). In this form, for a given fluid, a given structure, and a given flow velocity, everything on the right-hand side of Eq. (5.3) is constant, except for C_y, which represents the flow forces. Since the flow forces change with the angle of attack, α, then C_y is a function of α. In many cases, C_y is a linear function of α for small angles of attack and becomes a nonlinear function for larger angles of attack. Since C_y is a nonlinear function of α, and α is a function of $\dot{y}$ (Eq. (5.1)), then the equation of motion, which governs this system (Eq. (5.2)), is a nonlinear equation.

As a first step toward understanding the physics of the problem, we linearize the equation of motion about its original equilibrium position. Later on, in Sect. 5, we will consider the nonlinear equation. To linearize the equation of motion, we expand C_y using the Taylor series about a zero angle of attack, and only keep the first-order term. The expanded C_y reads as:

$$C_y = C_y\big|_{\alpha=0} + \frac{\partial C_y}{\partial \alpha}\bigg|_{\alpha=0} \alpha + O(\alpha^2), \tag{5.4}$$

where $O(\alpha^2)$ means all the terms of order 2 and higher. Note that in this equation, $C_y|_{\alpha=0}$ and $(\partial C_y/\partial \alpha)|_{\alpha=0}$ are constant values. Then we use Eqs. (5.3) and (5.4) to replace the right-hand side of Eq. (5.2):

$$m\ddot{y} + 2\zeta m\omega_n\dot{y} + ky = \frac{1}{2}\rho hU^2\left(C_y\big|_{\alpha=0} + \frac{\partial C_y}{\partial \alpha}\bigg|_{\alpha=0} \alpha\right). \tag{5.5}$$

Equation (5.5) seems to contain two unknowns, y on the left-hand side and α on the right-hand side. But, using Eq. (5.1), we can relate α to $\dot{y}$. For small angles of attack, we can linearize Eq. (5.1) by expanding it using the Taylor series, and keeping only the first (linear) term:

$$\alpha = \tan^{-1}\frac{\dot{y}}{U} = \frac{\dot{y}}{U} + O(\dot{y}^2), \tag{5.6}$$

where $O(\dot{y}^2)$ means all the terms of order 2 and higher.

Then we use Eq. (5.6), and replace α by $\dot{y}$ on the right-hand side of Eq. (5.5) to get:

$$m\ddot{y} + 2\zeta m\omega_n\dot{y} + ky = \frac{1}{2}\rho hU^2 C_y\big|_{\alpha=0} + \frac{1}{2}\rho hU^2\frac{\partial C_y}{\partial \alpha}\bigg|_{\alpha=0} \frac{\dot{y}}{U}, \tag{5.7}$$

and bring the new damping term (the term with a $\dot{y}$) from the right-hand side to the left-hand side:

$$m\frac{d^2y}{dt^2} + \left(2m\omega_n\zeta - \frac{1}{2}\rho hU\frac{\partial C_y}{\partial \alpha}\bigg|_{\alpha=0}\right)\frac{dy}{dt} + ky = \frac{1}{2}\rho hU^2 C_y\big|_{\alpha=0}, \tag{5.8}$$

or

$$m\ddot{y} + 2m\omega_n\left(\zeta - \frac{1}{2}\rho h U \frac{1}{2m\omega_n}\frac{\partial C_y}{\partial \alpha}\bigg|_{\alpha=0}\right)\dot{y} + ky = \frac{1}{2}\rho h U^2 C_y\big|_{\alpha=0}. \tag{5.9}$$

This is the equation of motion for a flexibly mounted structure with a nonuniform cross section and with 1 DOF in the transverse direction. The only flow-induced term on the left-hand side of this equation is the new damping term, which is subtracted from the structural damping. This suggests that at some point, the damping might become negative as U is varied. This is in fact the main mechanism of instability that results in galloping as we will discuss in Sect. 3.3. The flow-induced term on the right-hand side of the equation does not depend on y. It is a constant term, which implies that the system is under a static loading, and to respond to that, it will undergo a static deflection, besides any possible dynamic response.

3.2 *Dimensionless Equation of Motion for 1 DOF Galloping*

Before conducting any analysis, we make the equation of motion (Eq. (5.8)) dimensionless. We introduce two dimensionless parameters:

$$Y = \frac{y}{h} \text{ and } T = f_n t, \tag{5.10}$$

where $f_n = (1/2\pi)\sqrt{k/m}$ is the natural frequency of the system without the influence of the surrounding fluid in Hz. We plug these parameters into Eq. (5.8) to get:

$$m\frac{d^2(hY)}{d\left(\frac{T}{f_n}\right)^2} + \left(2m(2\pi f_n)\zeta - \frac{1}{2}\rho h U \frac{\partial C_y}{\partial \alpha}\bigg|_{\alpha=0}\right)\frac{d(hY)}{d\left(\frac{T}{f_n}\right)} + khY = \frac{1}{2}\rho h U^2 C_y\big|_{\alpha=0}, \tag{5.11}$$

or

$$mf_n^2 h\frac{d^2Y}{dT^2} + f_n h\left(2m(2\pi f_n)\zeta - \frac{1}{2}\rho h U \frac{\partial C_y}{\partial \alpha}\bigg|_{\alpha=0}\right)\frac{dY}{dT} + khY = \frac{1}{2}\rho h U^2 C_y\big|_{\alpha=0}, \tag{5.12}$$

and after dividing all terms by mf_n^2h:

$$\frac{d^2Y}{dT^2}+\frac{f_nh}{mf_n^2h}\left(2m(2\pi f_n)\zeta-\frac{1}{2}\rho hU\frac{\partial C_y}{\partial\alpha}\bigg|_{\alpha=0}\right)\frac{dY}{dT}+\frac{kh}{mf_n^2h}Y$$
$$=\frac{1}{2}\frac{\rho hU^2}{mf_n^2h}C_y\bigg|_{\alpha=0}, \tag{5.13}$$

and considering that $k/m=4\pi^2f_n^2$,

$$\frac{d^2Y}{dT^2}+\frac{1}{mf_n}\left(2m(2\pi f_n)\zeta-\frac{1}{2}\rho hU\frac{\partial C_y}{\partial\alpha}\bigg|_{\alpha=0}\right)\frac{dY}{dT}+\frac{4\pi^2f_n^2}{f_n^2}Y$$
$$=\frac{1}{2}\frac{\rho U^2}{mf_n^2}C_y\bigg|_{\alpha=0}, \tag{5.14}$$

or

$$\frac{d^2Y}{dT^2}+\left(\frac{2m(2\pi f_n)\zeta}{mf_n}-\frac{1}{2}\frac{\rho hU}{mf_n}\frac{\partial C_y}{\partial\alpha}\bigg|_{\alpha=0}\right)\frac{dY}{dT}+4\pi^2Y=\frac{1}{2}\frac{\rho U^2}{mf_n^2}C_y\bigg|_{\alpha=0}. \tag{5.15}$$

Then we multiply the second term of the damping coefficient, and the term on the hind side by h^2/h^2 to get:

$$\frac{d^2Y}{dT^2}+\left(4\pi\zeta-\frac{1}{2}\frac{\rho h}{mf_n}\frac{h^2}{h^2}\frac{\partial C_y}{\partial\alpha}\bigg|_{\alpha=0}\right)\frac{dY}{dT}+4\pi^2Y=\frac{1}{2}\frac{\rho U^2}{mf_n^2}\frac{h^2}{h^2}C_y\bigg|_{\alpha=0}, \tag{5.16}$$

which can be re-written as:

$$\frac{d^2Y}{dT^2}+\left(4\pi\zeta-\frac{1}{2}\frac{\rho h^2}{m}\frac{U}{f_nh}\frac{\partial C_y}{\partial\alpha}\bigg|_{\alpha=0}\right)\frac{dY}{dT}+4\pi^2Y=\frac{1}{2}\frac{\rho h^2}{m}\frac{U^2}{f_n^2h^2}C_y\bigg|_{\alpha=0}. \tag{5.17}$$

Here, we define $V_r=U/(f_nh)$ as a reduced velocity and $m_h^*=m/(\rho h^2)$ as a mass ratio. Note that while ρh^2 represents the volume of displaced fluid, it does not necessarily correspond to the exact volume of displaced fluid. For example, for a square cross section, ρh^2 is the actual volume of displaced fluid; however, for a triangular cross section, ρh^2 is twice the volume of displaced fluid.

The final version of the dimensionless linear equation then takes the following form:

$$\frac{d^2Y}{dT^2}+\left(4\pi\zeta-\frac{1}{2}\frac{V_r}{m_h^*}\frac{\partial C_y}{\partial\alpha}\bigg|_{\alpha=0}\right)\frac{dY}{dT}+4\pi^2Y=\frac{1}{2}\frac{V_r^2}{m_h^*}C_y\bigg|_{\alpha=0}, \tag{5.18}$$

in which the flow-induced damping as well as the constant flow-induced term on the right-hand side depend on both the mass ratio and the reduced velocity.

3.3 *The Criterion for Stability*

The solution of the equation of motion (Eq. (5.9)) consists of a homogenous solution, y_h (i.e., the response of the system with no external force), and a particular solution, y_p (i.e., the response of the system to the external force):

$$y = y_h + y_p, \tag{5.19}$$

which can be found to be

$$y = Ye^{-\zeta_T \omega_n t} \sin(\omega_d t + \phi) + \frac{\frac{1}{2}\rho h U^2 C_y|_{\alpha=0}}{k}, \tag{5.20}$$

where

$$\zeta_T = \zeta - \rho h U \frac{1}{4m\omega_n} \frac{\partial C_y}{\partial \alpha}\bigg|_{\alpha=0}. \tag{5.21}$$

A detailed discussion on homogeneous and particular solutions of these types of ODEs was given in Sect. 5 of Chap. 3. The homogenous solution is the same as the solution we obtained for a damped vibratory system in Chap. 3 (Eq. 5.47 in Chap. 3). Since the right-hand side of the equation of motion is a constant, the particular solution which should satisfy the equation including its right-hand side becomes a static response: $(\frac{1}{2}\rho h U^2 C_y|_{\alpha=0})/k$.

By looking at the form of the solution in Eq. (5.20), one might think that the system will always go to a static steady state: the first term is the transient response and will decay to zero, and the second term, which will take over during the steady state, is a static term. This scenario is correct only based on the assumption that the damping coefficient, ζ_T, is a positive number. However, ζ_T is the total damping ratio of the system, which includes the flow-induced damping, and it is not necessarily positive at all times. For some flow velocities where ζ_T is not positive, the structure exhibits a dynamic response. The onset of this dynamic response depends on the damping ratio (Eq. (5.21)), which can be re-arranged as:

$$\zeta_T = \zeta - \frac{U}{4\omega_n h} \frac{\rho h^2}{m} \frac{\partial C_y}{\partial \alpha}\bigg|_{\alpha=0}, \tag{5.22}$$

or using the dimensionless parameters, V_r and m_h^*, as:

$$\zeta_T = \zeta - \frac{1}{8\pi}\frac{V_r}{m_h^*}\left.\frac{\partial C_y}{\partial \alpha}\right|_{\alpha=0}. \tag{5.23}$$

The stability of the system depends on the sign of ζ_T: For a positive ζ_T, the overall damping of the system is positive and oscillations are damped. For a negative ζ_T, however, the system will have a negative damping, which means that oscillations are amplified, instead of being damped, and the amplitude of oscillations can grow to infinity—at least according to this linear model. This can be observed mathematically by looking at the solution for the equation of motion (Eq. (5.20)): If ζ_T is a negative number, then the homogeneous solution keeps growing exponentially with time and can become larger than the particular solution, because the particular solution is a constant. In that case, the solution will grow to infinity, which means an unstable solution. A positive ζ_T, however, results in exponential decay and the response will converge to the particular solution with time and therefore the system is stable.

Since the structural damping, ζ, is always positive, then the sign of the total damping, ζ_T, depends on the sign of the flow-induced damping. In the flow-induced damping term, $1/(8\pi)$, V_r and m_h^* are positive, then the stability depends on the sign of $(\partial C_y/\partial \alpha)|_{\alpha=0}$, which is the slope of the $C_y - \alpha$ curve at $\alpha = 0$. If $(\partial C_y/\partial \alpha)|_{\alpha=0}$ is a negative number, then the total damping coefficient is a positive number and therefore the system is stable. If, however, $(\partial C_y/\partial \alpha)|_{\alpha=0}$ is positive, the second term in the total damping will become negative and could be larger than the structural damping, which implies a negative total damping, and therefore instability. Then for instability we should have:

$$\left.\frac{\partial C_y}{\partial \alpha}\right|_{\alpha=0} > 0. \tag{5.24}$$

Furthermore, the absolute value of the second term in the total damping ratio (Eq. (5.21)) must be larger than ζ.

3.4 *The Onset of Instability*

The onset of instability is when ζ_T becomes negative. Then to find the critical flow velocity at which instability occurs, we should find the reduced velocity at which $\zeta_T = 0$, which means:

$$\frac{1}{8\pi}\frac{V_{r,cr}}{m_h^*}\left.\frac{\partial C_y}{\partial \alpha}\right|_{\alpha=0} = \zeta, \tag{5.25}$$

or

$$V_{r,cr} = \frac{8\pi m_h^* \zeta}{\left.\frac{\partial C_y}{\partial \alpha}\right|_{\alpha=0}}. \tag{5.26}$$

In the numerator of this critical reduced velocity, the term $m_h^* \zeta$ shows up. This is what we observed in our discussion on VIV in Chap. 4 as well, and we called it a mass-damping parameter (Sect. 5). It seems that the mass-damping parameter has some relevance here as well: a larger mass-damping parameter results in a larger critical value for the onset of instability, for a given geometry. (A given geometry means a constant value for the slope—i.e., the denominator of Eq. (5.26).)

3.5 *The Flow Forces in Terms of C_L and C_D*

The flow force coefficient, C_y, represents the flow forces in the vertical direction (perpendicular to the incoming flow velocity) only, and can be measured (or calculated numerically) as the oscillations are ongoing. For bluff bodies with different cross sections, measurements of lift and drag coefficients do exist for cases when the structure is fixed in flow. In these cases, the lift and drag coefficients are simply coefficients of flow forces perpendicular to and in line with the incoming flow, respectively. In the case we consider here, however, the lift and drag coefficients would be perpendicular to and in line with the relative flow velocity, because of the oscillations of the structure. If we relate C_y to C_L and C_D, then we can use the existing data in the literature, where C_L and C_D have been measured versus α, in solving the equation of motion of our system.

Following the free-body-diagram of Fig. 5.3, we can relate F_y to F_L and F_D as:

$$F_y = -F_L \cos\alpha - F_D \sin\alpha. \tag{5.27}$$

Since lift and drag are forces perpendicular to and in parallel with the relative velocity, U_{rel}, respectively, we can write them as

$$F_L = \frac{1}{2}\rho h U_{rel}^2 C_L, \tag{5.28}$$

and

$$F_D = \frac{1}{2}\rho h U_{rel}^2 C_D, \tag{5.29}$$

where we use U_{rel} in the coefficient. From Eq. (5.3):

$$F_y = \frac{1}{2}\rho h U^2 C_y, \tag{5.30}$$

and therefore we can write Eq. (5.27) as:

$$\frac{1}{2}\rho h U^2 C_y = -\frac{1}{2}\rho h U_{rel}^2 (C_L \cos\alpha + C_D \sin\alpha), \tag{5.31}$$

which leads to a relation for C_y in terms of C_L and C_D:

$$C_y = -\left(\frac{U_{rel}}{U}\right)^2 (C_L \cos\alpha + C_D \sin\alpha). \tag{5.32}$$

Equation (5.32) is a nonlinear relation for C_y versus α, since it includes terms with $\sin\alpha$ and $\cos\alpha$, and α can be large. Since in this section we are considering a linear form of the problem, we will linearize this equation about a zero angle of attack.

First of all, we note from Eq. (5.32) that at $\alpha = 0$,

$$C_y\big|_{\alpha=0} = -C_L\big|_{\alpha=0}. \tag{5.33}$$

Then we use Eq. (5.32) and find the derivative of C_y with respect to α at $\alpha = 0$:

$$\left.\frac{\partial C_y}{\partial \alpha}\right|_{\alpha=0} = -\left(\frac{U_{rel}}{U}\right)^2 \left.\left(\frac{\partial C_L}{\partial \alpha}\cos\alpha - C_L \sin\alpha + \frac{\partial C_D}{\partial \alpha}\sin\alpha + C_D \cos\alpha\right)\right|_{\alpha=0}, \tag{5.34}$$

or

$$\left.\frac{\partial C_y}{\partial \alpha}\right|_{\alpha=0} = -\left(\frac{U_{rel}}{U}\right)^2 \left.\left(\frac{\partial C_L}{\partial \alpha} + C_D\right)\right|_{\alpha=0}. \tag{5.35}$$

We note that at $\alpha = 0$,

$$U_{rel}\big|_{\alpha=0} = U, \tag{5.36}$$

and therefore Eq. (5.35) becomes:

$$\left.\frac{\partial C_y}{\partial \alpha}\right|_{\alpha=0} = -\left.\left(\frac{\partial C_L}{\partial \alpha} + C_D\right)\right|_{\alpha=0}. \tag{5.37}$$

Then by plugging Equations (5.33) and (5.37) into Eq. (5.4):

$$C_y = -C_L\big|_{\alpha=0} - \left.\left(\frac{\partial C_L}{\partial \alpha} + C_D\right)\right|_{\alpha=0} \alpha. \tag{5.38}$$

Since, according to Eq. (5.24), to observe instability we need to have:

$$\left.\frac{\partial C_y}{\partial \alpha}\right|_{\alpha=0} > 0, \tag{5.39}$$

we can re-write this criterion in terms of C_L and C_D using Eq. (5.37) as:

$$\left.\left(\frac{\partial C_L}{\partial \alpha} + C_D\right)\right|_{\alpha=0} < 0. \tag{5.40}$$

Then we can use Eq. (5.40) to predict possible galloping of a 1 DOF prism placed in flow based on the existing measurements of its lift and drag coefficients versus the angle of attack.

3.6 *Linear Stability Analysis*

The linear equation of a galloping system (Eq. 5.9) is a sample of linear equations of motion that are observed extensively in dynamical systems, and in FSI systems. In these systems, we have one (or multiple) second-order linear ODEs whose coefficients are functions of the system parameters and could change with changing an independent variable—the flow velocity as an example. These equations, in their general form, are written as:

$$m(U)\ddot{y} + c(U)\dot{y} + k(U)y = 0, \tag{5.41}$$

in which m, c, and k are all functions of an independent variable, U. The solution of this equation has the form of:

$$y = Ye^{\lambda t}. \tag{5.42}$$

If we plug this general form of solution into the equation, we get:

$$m(U)\lambda^2 + c(U)\lambda + k(U) = 0. \tag{5.43}$$

For any given value of U, values of λ can be found from Eq. (5.43). These λ values are called the system's eigenvalues, and in general are complex numbers, which can be written as:

$$\lambda = \lambda_R \pm i\lambda_I, \tag{5.44}$$

in which $i = \sqrt{-1}$.

If we plug back Eq. (5.44) into Eq. (5.42), the exponential part of the solution will find the following form:

$$e^{(\lambda_R+i\lambda_I)t} = e^{\lambda_R t}e^{i\lambda_I t}, \tag{5.45}$$

in which, the first term, $e^{\lambda_R t}$, corresponds to an exponential decay for negative λ_R or an exponential growth for positive λ_R. An exponential decay implies a stable solution because if the system is disturbed from its equilibrium, it goes back to it. An exponential growth means an unstable solution, because a small disturbance of the system results in a solution that grows exponentially, without any limit. Then, λ_R acts as a damping, but with a negative sign, which we refer to as a "negative damping". The second term, $e^{i\lambda_I t}$, can be written in terms of $\sin\lambda_I t$ and $\cos\lambda_I t$ using Euler's formula:

$$e^{i\lambda_I t} = \cos\lambda_I t + i\sin\lambda_I t. \tag{5.46}$$

Then λ_I represents the frequency of the system's response.

Clearly, λ_R and λ_I change as U is increased. Usually, the system is stable at zero flow velocity. Then as the flow velocity is increased, at some critical flow velocity, the system could become unstable. This is investigated by plotting λ_R and λ_I versus U or by plotting λ_I versus λ_R in a plot called the Argand diagram.

3.7 *The Eigenvalue Plots and the Argand Diagram*

The Argand diagram summarizes how λ_R and λ_I vary with flow velocity by plotting λ_I versus λ_R. This can also be done by plotting λ_R and λ_I versus the flow velocity. Figure 5.5 shows samples of these plots for three commonly observed instabilities. In every case, an independent variable is changed (flow velocity in the case of our problem here) and the corresponding values of λ_R and λ_I at each flow velocity are plotted. In all plots, at $U = 0$, the positive value of λ_I corresponds to the natural frequency of the dry system (i.e., only the structure, without any flow or fluid around it) and the negative λ_R corresponds to the structural damping ratio. Note that the negative λ_I at $U = 0$ does not have any physical significance, since there is no negative natural frequency for a mass-spring-damper system. This negative λ_I is only a byproduct of the mathematical solution of the eigenvalue problem. Here, we discuss the three scenarios with some details. Figure 5.5 shows the Argand diagrams (left) as well as the plots of imaginary and real parts of the eigenvalue for these cases.

(i) *Buckling*

In plots of the first row of Fig. 5.5, as the independent variable, U, is increased, λ_I decreases and becomes zero at a critical value of U, referred to as U_B. At $U = U_B$, $\lambda_I = 0$, which means that the system has zero natural frequency (i.e., zero stiffness),

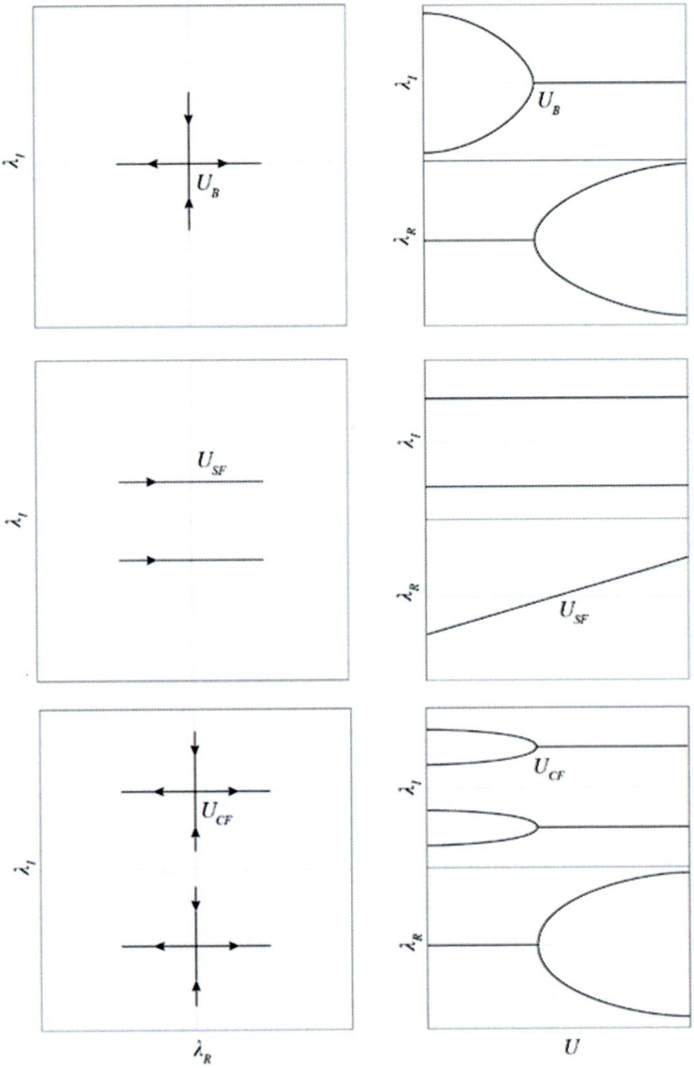

Fig. 5.5 Schematic of Argand diagrams and eigenvalue plots corresponding to different instabilities: First row: buckling; second row: single-mode flutter; third row: coupled-mode flutter

and therefore it buckles. This is called a divergence (buckling) instability, and U_B is called the critical flow velocity for buckling. Since this is a completely static response, the damping ratio, represented by λ_R, has no significance.

(ii) *Single-mode Flutter*

In plots of the second row of Fig. 5.5, as the independent variable, U, is increased, λ_R, which is negative initially, becomes positive at a critical U, referred to as U_{SF}. A positive λ_R corresponds to a negative damping, and as a result, the solution becomes unstable. These values of negative damping correspond to positive values of λ_I (i.e., positive natural frequency), which means that the system oscillates with a frequency and since the solution is unstable, the amplitude of its response increases exponentially with time. This is called a *single-mode flutter.*

(iii) *Coupled-mode Flutter*

In plots of the third row of Fig. 5.5, as the independent variable, U, is increased, λ_I values corresponding to two different modes of the system approach each other and at a critical value of U, referred to as U_{CF}, they merge. At the same value of the independent variable, a positive value of λ_R is observed, which means that the system does have a negative damping. Again, the system oscillates with a nonzero frequency and since the solution is unstable, the amplitude of these oscillations increases exponentially. Since two modes are involved in this instability, it is called a *coupled-mode flutter*. Once coupled-mode flutter occurs, the system starts oscillating, and the frequency of oscillations in both degrees of freedom will be the same and equal to the flutter frequency. Obviously, the system has to have at least 2 degrees of freedom to be susceptible to coupled-mode flutter.

3.8 Linear Stability Analysis for Galloping

In order to study the instability of a galloping system, we consider the dimensionless equation of motion (Eq. 5.18):

$$\frac{d^2Y}{dT^2} + \left(4\pi\zeta - \frac{1}{2}\frac{V_r}{m_h^*}\frac{\partial C_y}{\partial \alpha}\bigg|_{\alpha=0}\right)\frac{dY}{dT} + 4\pi^2 Y = \frac{1}{2}\frac{V_r^2}{m_h^*}C_y\big|_{\alpha=0}. \tag{5.47}$$

The criterion for instability is the same as the criterion discussed in Sect. 3.3: When the damping term becomes negative, instability is observed. Then the instability occurs when:

$$4\pi\zeta - \frac{1}{2}\frac{V_r}{m_h^*}\frac{\partial C_y}{\partial \alpha}\bigg|_{\alpha=0} = 0, \tag{5.48}$$

which gives a critical value for the reduced velocity as:

$$V_{r,cr} = \frac{8\pi m_h^*\zeta}{\left.\frac{\partial C_y}{\partial \alpha}\right|_{\alpha=0}}, \tag{5.49}$$

which is the same value as what we found in Eq. (5.26).

We can find this critical point for instability numerically as well by varying V_r from 0 to higher values, calculating the system's eigenvalues for each value of V_r, and plotting the real and imaginary parts of these eigenvalues versus the reduced velocity. To do so, first we have to consider a specific geometry for the cross section of the bluff body, placed at a specific angle. For example, we consider a square cross section placed at a 90° angle, such that two of its sides are perpendicular to the incoming flow. By considering a specific geometry and a specific angle of attack, we obtain $C_y|_{\alpha=0}$. For our case, this will be $C_y|_{\alpha=0} = 2.7$. We further need the values of m_h^* and ζ. To calculate $m_h^* = m/(\rho h^2)$, we need the mass per unit length of the structure, m, which we assume to be $m = 1$ kg/m, the side of the square, h, which we assume to be $h = 0.1$ m, and the density of fluid, which we assume to be air, whose density is $\rho = 1.124\ kg/m^3$. These values lead to $m_h^* = 89$. We further assume a small structural damping ratio: $\zeta = 0.01$. We also note that our independent variable, V_r, is related to the incoming flow velocity. To find this relation, we need the system's natural frequency, f_n, which we assume to be $f_n = 4$ Hz. Then the reduced velocity, V_r, is related to U as $V_r = U/(f_n h) = U/(4 \times 0.1) = 2.5U$.

With these system parameters, the critical reduced velocity for the onset of galloping can be found from Eq. (5.49) to be $V_{r,cr} = 8.3$. This critical value can also be found by looking at the plot of the real parts and imaginary parts of the eigenvalues of the system of Eq. (5.47) as V_r is varied. This plot is shown in Fig. 5.6.

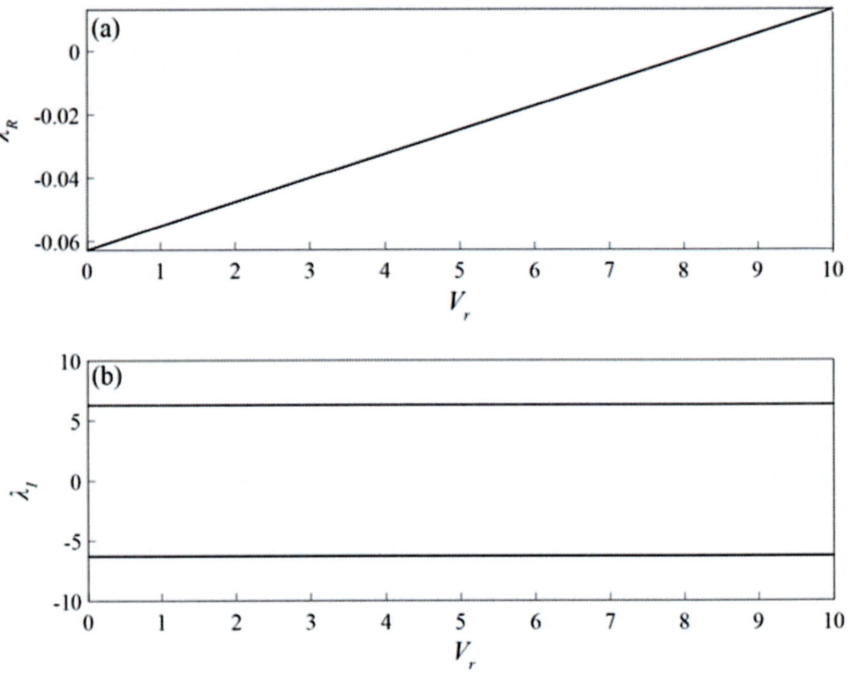

Fig. 5.6 (**a**) The real and (**b**) imaginary parts of the eigenvalues of a square prism placed in flow, indicating the onset of instability at $V_{r,cr} = 8.3$

The real part of the eigenvalue, which is proportional to a negative damping, is initially negative, indicating a positive damping, and a stable solution. As the reduced velocity is increased, the real part of the eigenvalue increases linearly, and it crosses zero at $V_{r,cr} = 8.3$, which is the same critical value for the onset of galloping that we found using Eq. (5.49). Note that the imaginary part of the eigenvalue remains constant for all values of reduced velocity, which implies that changing the reduced velocity does not change the dimensionless natural frequency of the system, as is also obvious in Eq. (5.47). The plots of Fig. 5.6 indicate that the system starts oscillating at $V_{r,cr} = 8.3$, with a dimensionless frequency of 2π. These plots correspond to what we saw in Fig. 5.5 as a typical plot for a single-mode flutter instability.

3.9 On the Quasi-Steady Assumption and the Differences Between VIV and Galloping

The values of C_L and C_D, which can be found in the literature, are based on experiments conducted when the body is fixed at a given angle of attack. Then in Eq. (5.40) we use the results of a series of static experiments to predict a dynamic response. This is based on the assumption that the flow forces that act on the oscillating structure at every instant of time are equal to flow forces that act on the structure when it is held static at that instant of time. This assumption also implies that the flow forces at every instant of time are independent from the motion of the structure prior to that time. This is called a *quasi-steady* assumption and is acceptable only if the oscillations of the structure are very slow—very low-frequency oscillations.

Galloping is considered to be a large-amplitude and low-frequency oscillation. This is a low-frequency response, where "low" is in comparison with the shedding frequency as opposed to a VIV response in which the oscillation and shedding frequencies are equal. The low-frequency galloping oscillation implies that the shedding is the result of the presence of the bluff body in flow and not the cause of its oscillations. Several vortices can be shed during one cycle of a galloping response. The large-amplitude oscillation is usually in comparison with the amplitude of the VIV response as well, which is typically around one cylinder diameter. In galloping, amplitudes up to several times the side of the cross section can be observed.

While the discussions in this section were focused on galloping of a bluff body, they are also valid for the cases where a streamlined body (such as an airfoil) is placed in flow. If the airfoil is asymmetric by design or if it is placed at an angle with respect to the incoming flow, a nonzero mean lift force will act on it (besides the nonzero mean drag force) and the same scenario as what we discussed for galloping of a bluff body could occur, as long as the airfoil is free to oscillate in 1 DOF only.

A bluff body with 2 degrees of freedom placed in flow could undergo 2D galloping as well. We will discuss instabilities of 2 DOF systems with a focus on a streamlined structure in the following section.

4 A Streamlined Structure in Flow

In this section, we consider flow-induced instabilities of a structure with a noncircular cross section and with two degrees of freedom placed in flow. We assume that the two degrees of freedom are in the transverse, y, and torsional, θ, directions (Fig. 5.7). While the general discussion would also be valid for a structure with a bluff cross section, here we focus on a structure with a streamlined cross section (i.e., an airfoil). Figure 5.7 shows an airfoil with two degrees of freedom placed in flow. The airfoil has a chord length of c, and is supported by two springs: one linear spring with a spring contact k, and one torsional spring with a spring constant K_t. From this schematic, it is clear that the incoming flow exerts a distributed force along the length of the airfoil. It is also clear that the airfoil's response will depend on the location where the springs are attached to the airfoil, and its center of gravity. In Fig. 5.7, A corresponds to the aerodynamic center of the airfoil, which is the location along its chord where an equivalent single flow force replaces the

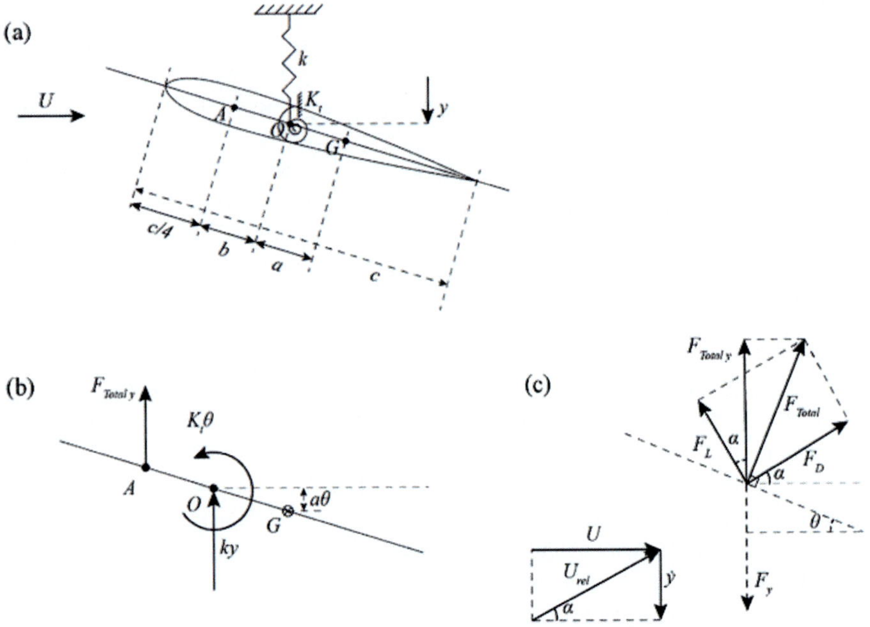

Fig. 5.7 (**a**) An airfoil with two degrees of freedom placed in flow (**b**) the forces that act on the airfoil, and (**c**) the details of how the force component in the vertical direction depends on F_L and F_D

distributed flow force acting on the airfoil (this is generally assumed to be located at a distance $c/4$ from the airfoil's leading edge, which is where the flow sees the airfoil first), G corresponds to the center of gravity, and O to the elastic axis, which is the point where the spring is attached to the airfoil (this is normally considered as the point about which an applied force would not produce any torsion, and an applied torque would not result in any linear displacement).

4.1 The Equations of Motion for a 2 DOF Airfoil in Flow

To derive the equations of motion of the system, we use the free body diagram of Fig. 5.7b. Note that we still assume y is positive downward. Then in the y-direction, we can write

$$\sum F_y = ma, \tag{5.50}$$

and then

$$-F_{Total,y} - ky = m(\ddot{y} + a\ddot{\theta}), \tag{5.51}$$

where $F_{Total,y}$ is the total force that acts on the airfoil in the upward vertical direction, and can be written in terms of F_L and F_D as (Fig. 5.7c):

$$F_{Total,y} = F_D \sin\alpha + F_L \cos\alpha. \tag{5.52}$$

For a small angle of attack:

$$F_{Total,y} = F_D\alpha + F_L, \tag{5.53}$$

and if we assume $F_D \ll F_L$, which is a logical assumption for a streamlined body:

$$F_{Total,y} = F_L, \tag{5.54}$$

and therefore Eq. (5.51) can be written as

$$-F_L - ky = m(\ddot{y} + a\ddot{\theta}). \tag{5.55}$$

Note that we can assume the drag force is much smaller than the lift force, only when the airfoil stays at small angles of attack, which is a fine assumption here, as we are deriving the linear equations of motion for this system.

Following the free body diagram of Fig. 5.7b, in the torsional direction we write:

$$\sum T_G = J_G\ddot{\theta}, \tag{5.56}$$

where the left-hand side is the summation of all the external torques that act on the body about point G, and J_G is the mass moment of inertia of the airfoil about its center of gravity. Therefore,

$$+kya - K_t\theta + F_L(a+b) = J_G\ddot{\theta}. \tag{5.57}$$

Or we can write the torques about point O and find:

$$\sum T_O = J_O\ddot{\theta}, \tag{5.58}$$

where J_O is the mass moment of inertia of the airfoil about its elastic axis, which results in

$$-K_t\theta + F_L b = m\ddot{y}a + (J_G + ma^2)\ddot{\theta}. \tag{5.59}$$

Equation (5.59) can be found if we multiply Eq. (5.55) by a and add it to Eq. (5.57), and therefore Eqs. (5.57) and (5.59) are not independent and we should use one of them—either one. Here, we use Eqs. (5.55) and (5.59) as the equations of motion:

$$\begin{aligned} &m\ddot{y} + ma\ddot{\theta} + ky = -F_L, \\ &(J_G + ma^2)\ddot{\theta} + ma\ddot{y} + K_t\theta = F_L b, \end{aligned} \tag{5.60}$$

which can be written in a matrix form as:

$$\begin{bmatrix} m & ma \\ ma & J_G + ma^2 \end{bmatrix}\begin{Bmatrix} \ddot{y} \\ \ddot{\theta} \end{Bmatrix} + \begin{bmatrix} k & 0 \\ 0 & K_t \end{bmatrix}\begin{Bmatrix} y \\ \theta \end{Bmatrix} = \begin{Bmatrix} -F_L \\ F_L b \end{Bmatrix}, \tag{5.61}$$

where the lift force can be written as:

$$F_L = \frac{1}{2}\rho c U^2 C_L. \tag{5.62}$$

We know that C_L is a nonlinear function of the angle of attack, and therefore we can use the Taylor series to expand it about $\alpha = 0$:

$$C_L = C_L|_{\alpha=0} + \left.\frac{\partial C_L}{\partial \alpha}\right|_{\alpha=0} \alpha + O(\alpha^2). \tag{5.63}$$

Then we plug Eqs. (5.62) and (5.63) back into Eq. (5.60) to get:

$$\begin{aligned} m\ddot{y} + ma\ddot{\theta} + ky &= -\frac{1}{2}\rho c U^2 \left[C_L|_{\alpha=0} + \left.\frac{\partial C_L}{\partial \alpha}\right|_{\alpha=0} \alpha \right], \\ (J_G + ma^2)\ddot{\theta} + ma\ddot{y} + K_t\theta &= \frac{1}{2}\rho c U^2 \left[C_L|_{\alpha=0} + \left.\frac{\partial C_L}{\partial \alpha}\right|_{\alpha=0} \alpha \right] b. \end{aligned} \tag{5.64}$$

In this set of equations, we have both α (the angle of attack) and θ (the angle of rotation). Note that these two angles are not necessarily the same: One, α, depends on the direction of the relative flow velocity, and the other, θ, on the orientation of the airfoil (Fig. 5.7). Here, however, we assume that for a linear system, when the angles are small, α and θ are the same: i.e., the angle of attack is the same as the angle of rotation. Then we replace α by θ to get:

$$\begin{aligned} m\ddot{y} + ma\ddot{\theta} + ky &= -\frac{1}{2}\rho c U^2 C_L|_{\alpha=0} - \frac{1}{2}\rho c U^2 \left.\frac{\partial C_L}{\partial \alpha}\right|_{\alpha=0} \theta, \\ (J_G + ma^2)\ddot{\theta} + ma\ddot{y} + K_t\theta &= \frac{1}{2}\rho c U^2 b C_L|_{\alpha=0} + \frac{1}{2}\rho c U^2 \left.\frac{\partial C_L}{\partial \alpha}\right|_{\alpha=0} b\theta, \end{aligned} \tag{5.65}$$

and take all the unknowns to the left-hand side to find the final form of the equations of motion:

$$\begin{aligned} m\ddot{y} + ma\ddot{\theta} + ky + \frac{1}{2}\rho c U^2 \left.\frac{\partial C_L}{\partial \alpha}\right|_{\alpha=0} \theta &= -\frac{1}{2}\rho c U^2 C_L|_{\alpha=0}, \\ (J_G + ma^2)\ddot{\theta} + ma\ddot{y} + \left(K_t - \frac{1}{2}\rho c U^2 \left.\frac{\partial C_L}{\partial \alpha}\right|_{\alpha=0} b \right)\theta &= \frac{1}{2}\rho c U^2 b C_L|_{\alpha=0}, \end{aligned} \tag{5.66}$$

which can be written in a matrix form as:

$$\begin{aligned} &\begin{bmatrix} m & ma \\ ma & J_G + ma^2 \end{bmatrix} \begin{Bmatrix} \ddot{y} \\ \ddot{\theta} \end{Bmatrix} + \begin{bmatrix} k & \frac{1}{2}\rho c U^2 \left.\frac{\partial C_L}{\partial \alpha}\right|_{\alpha=0} \\ 0 & K_t - \frac{1}{2}\rho c U^2 \left.\frac{\partial C_L}{\partial \alpha}\right|_{\alpha=0} b \end{bmatrix} \begin{Bmatrix} y \\ \theta \end{Bmatrix} \\ &= \begin{Bmatrix} -\frac{1}{2}\rho c U^2 C_L|_{\alpha=0} \\ \frac{1}{2}\rho c U^2 b C_L|_{\alpha=0} \end{Bmatrix}, \end{aligned} \tag{5.67}$$

or

$$\mathbf{M}\,\ddot{\mathbf{Y}} + \mathbf{K}\,\mathbf{Y} = \mathbf{F}, \tag{5.68}$$

where

$$\mathbf{M} = \begin{bmatrix} m & ma \\ ma & J_G + ma^2 \end{bmatrix}, \tag{5.69}$$

$$\mathbf{K} = \begin{bmatrix} k & \frac{1}{2}\rho c U^2 \frac{\partial C_L}{\partial \alpha}\Big|_{\alpha=0} \\ 0 & K_t - \frac{1}{2}\rho c U^2 \frac{\partial C_L}{\partial \alpha}\Big|_{\alpha=0} b \end{bmatrix}, \tag{5.70}$$

and

$$\mathbf{F} = \begin{Bmatrix} -\frac{1}{2}\rho c U^2 C_L|_{\alpha=0} \\ \frac{1}{2}\rho c U^2 b C_L|_{\alpha=0} \end{Bmatrix}. \tag{5.71}$$

Equation (5.67) has the same form as Eq. (5.8), but instead of only 1 equation, we have two coupled equations, and the solutions have homogenous and particular parts, as we will see in the following section. Another difference is that in Eq. (5.67) there is no damping—neither structural damping (which we assumed to be zero) nor flow-induced damping. This suggests that if we observe an instability in this system, the mechanism of instability cannot be the same as what we saw for galloping, since there is no flow-induced damping in the equation to become negative as the flow velocity is increased,

4.2 *Solving the Equations of Motion and the Possible Instabilities of the System*

Similar to the equation of motion of a 1 DOF system that we discussed in Sect. 3, Eq. (5.67) represents a set of nonhomogeneous equations and has a nonzero right-hand side, which is a constant vector. The solution, therefore, will have a homogeneous and a particular part. Similar to the 1 DOF system, the particular solution of the system will be a constant vector.

To find the homogeneous solution of the equations of motion (i.e., the solution when the right-hand side of Eqs. (5.67) is zero), we consider the general solutions to be:

$$\begin{aligned} y &= A_y e^{\lambda t}, \\ \theta &= A_\theta e^{\lambda t}, \end{aligned} \tag{5.72}$$

and plug them into Eqs. (5.67) with a zero right-hand side to get:

$$\begin{bmatrix} m\lambda^2 + k & ma\lambda^2 + \frac{1}{2}\rho c U^2 \frac{\partial C_L}{\partial \alpha}\Big|_{\alpha=0} \\ ma\lambda^2 & (J_G + ma^2)\lambda^2 + K_t - \frac{1}{2}\rho c U^2 \frac{\partial C_L}{\partial \alpha}\Big|_{\alpha=0} b \end{bmatrix} \begin{Bmatrix} A_y \\ A_\theta \end{Bmatrix} = \begin{Bmatrix} 0 \\ 0 \end{Bmatrix}. \tag{5.73}$$

Eq. (5.73) constitutes an eigenvalue problem in which there are three unknowns: λ, A_y, and A_θ. Clearly, we cannot solve this set of two equations and find all three unknowns. The main unknown that is of significant interest for us is λ. By solving an eigenvalue problem, we find the values of λ and use them to discuss the state of the system. The trivial solution for this set of equations is $A_y = A_\theta = 0$, which satisfies the equations, but means that there is no nonzero response (i.e., the system stays at its original static position at all times). Since we are interested in observing a nonzero response of the system, we should find the nontrivial solutions.

To obtain the nontrivial solutions, the determinant of the 2×2 matrix of Eq. (5.73) must be equal to zero, which results in an algebraic equation:

$$\begin{aligned} &[m(J_G + ma^2) - (ma)^2]\lambda^4 \\ &+ \left[k(J_G + ma^2) + m\left(K_t - \frac{1}{2}\rho c U^2 \frac{\partial C_L}{\partial \alpha}\Big|_{\alpha=0} b\right) - ma\frac{1}{2}\rho c U^2 \frac{\partial C_L}{\partial \alpha}\Big|_{\alpha=0}\right]\lambda^2 \\ &+ \left[k\left(K_t - \frac{1}{2}\rho c U^2 \frac{\partial C_L}{\partial \alpha}\Big|_{\alpha=0} b\right)\right] = 0. \end{aligned} \tag{5.74}$$

Since all the parameters inside all the brackets in Eq. (5.74) are known for a given system, Eq. (5.74) can be written as:

$$C_4\lambda^4 + C_2\lambda^2 + C_0 = 0, \tag{5.75}$$

where C_0, C_2 and C_4 are constants. Equation (5.75) can be solved for λ, which results in 4 complex values for λ:

$$\lambda = \lambda_R + i\lambda_I. \tag{5.76}$$

We should note that the constants C_0, C_2, and C_4 change with flow velocity, U, which means that for every given flow velocity, we will calculate a new set of λ values. We can then plot the Argand diagrams or plots of the real part and imaginary parts of λ versus U to study the stability of the system. Two samples of these plots are shown in Fig. 5.8 (for a positive a), and in Fig. 5.9 (for a negative a). The other parameters are the same for both plots: $c = 1$ m, $b = 0.1$ m, $\rho = 1.124$ kg/m^3, $m = 10$ kg/m, $J_G = 4$ kg m^2/m, $k = 17$ N/m, and $K_t = 20$ N.m/rad. The horizontal

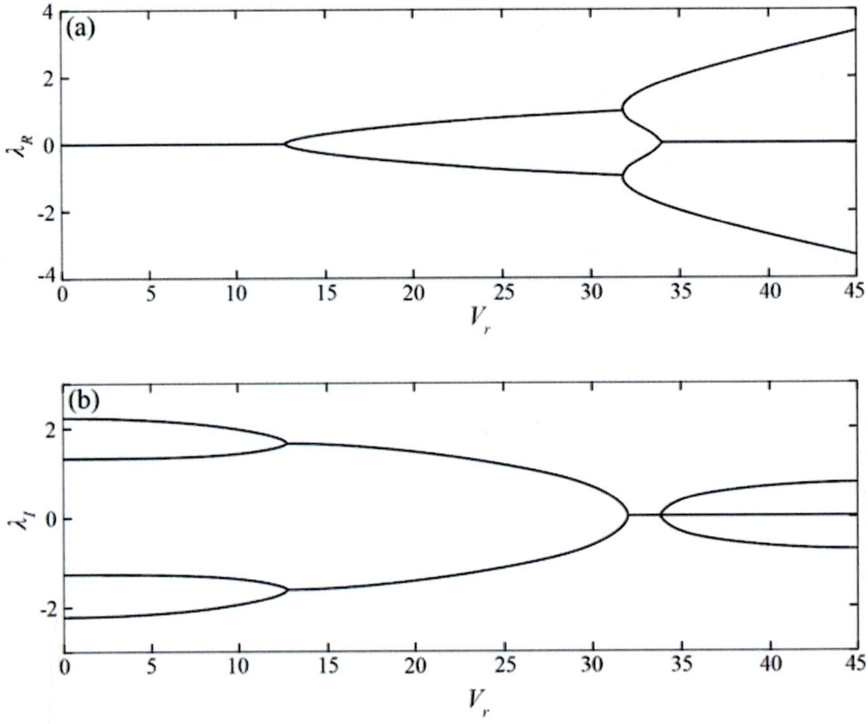

Fig. 5.8 (**a**) Real and (**b**) imaginary parts of the eigenvalues versus V_r for a 2 DOF airfoil with $a^* = 0.1$

axis of these plots represents the reduced velocity defined as $V_r = U/(f_n c)$, where $f_n = (1/2\pi)\sqrt{k/m}$. We also use a dimensionless value for a, defied as $a^* = a/c$. Clearly, the two plots correspond to two different types of instability:

For the same airfoil, when $a^* = 0.1$ (Fig. 5.8), the two frequencies of the system (corresponding to its transverse and torsional modes) move toward each other as the reduced velocity is increased and merge at a critical reduced velocity of $V_{r\_CF} = 12.4$. At the same critical reduced velocity, a negative damping is also observed in the system. This behavior corresponds to the coupled-mode flutter instability that we discussed in Sect. 3.7 (Fig. 5.5).

When $a^* = -0.1$ (Fig. 5.9), the frequency that corresponds to the transverse oscillations decreases as the reduced velocity is increased and reaches zero at a critical reduced velocity of $V_{r\_B} = 36.2$, suggesting that the airfoil buckles at this critical reduced velocity.

In general, when $a^* < 0$ (i.e., when the center of gravity is ahead of the elastic axis), the airfoil becomes unstable by buckling, and when $a^* > 0$ (i.e., when the center of gravity is behind the elastic axis), it undergoes coupled-mode flutter. Figure 5.10 summarizes how the critical values for the onset of buckling and coupled-mode flutter change with a^*. The critical value for buckling is independent

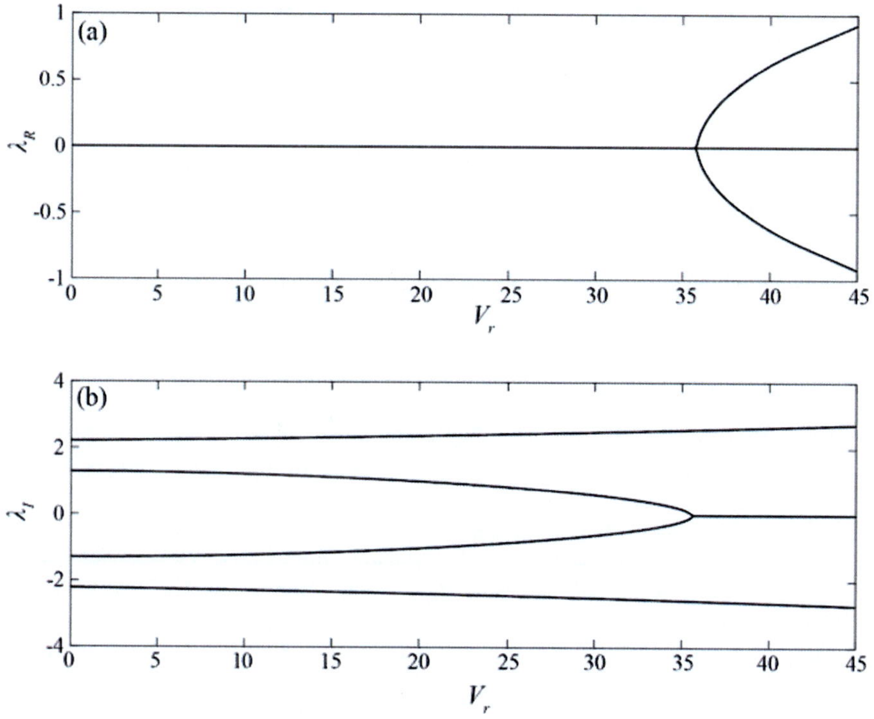

Fig. 5.9 (**a**) Real and (**b**) imaginary parts of the eigenvalues versus V_r for a 2 DOF airfoil with $a^* = -0.1$

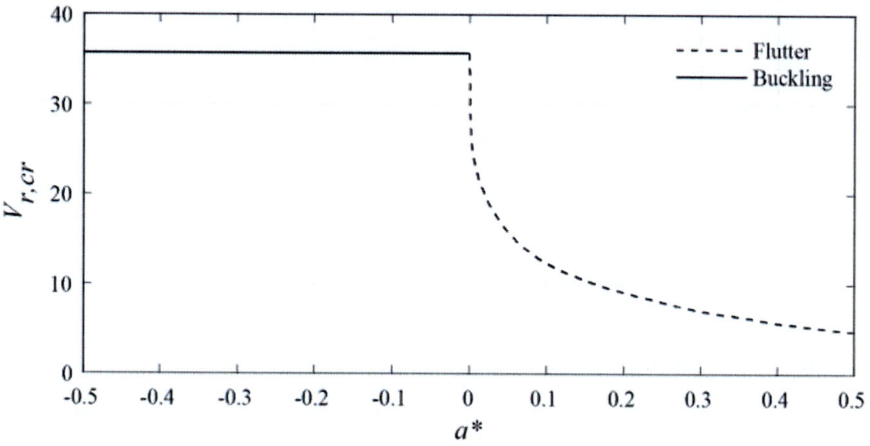

Fig. 5.10 Critical values for buckling and coupled mode flutter versus a^*

from a^*. This is expected, since in the equation of motion (Eq. (5.67)), the stiffness term and the external force are not functions of a^*, and since buckling is a static instability, only static terms of the equations of motion contribute to its onset. The critical reduced velocity for flutter decreases as a^* increases. This implies that the farther the center of gravity gets from the elastic center, it is more likely for the structure to undergo flutter.

5 Nonlinear Galloping

Consider the 1 DOF equation of motion for galloping that we discussed in Sect. 3.1:

$$m\ddot{y} + 2\zeta m\omega_n\dot{y} + ky = \frac{1}{2}\rho hU^2C_y. \tag{5.77}$$

As we saw in Sect. 3.1, C_y is a nonlinear function of α (Eq. 5.4) and α itself is a nonlinear function of $\dot{y}/U$ (Eq. 5.6). Therefore, the right-hand side of Eq. (5.77) is a nonlinear function of $\dot{y}/U$, and thus Eq. (5.77) is a nonlinear equation. In our linear analysis of this system in Sect. 3.1, we used the Taylor series to expand C_y versus $\dot{y}/U$ and kept only the linear term of the expansion (Eq. (5.6)). If we keep the higher order terms in our expansion, then we will have a nonlinear model for galloping of a 1 DOF system.

Why is a nonlinear model helpful? Among other benefits, a nonlinear model gives us the amplitude and frequency of the response after the first point of instability. For a structure that is placed in flow, before the first point of instability, there is no displacement and the linear assumptions of small angle of attack and small displacement are valid. After the first point of instability, linear assumptions are no longer valid and therefore a linear model cannot give us reliable information.

The nonlinear function of C_y versus $\dot{y}/U$ is found by curve fitting of experimental data for a given cross section. Figure 5.11 shows some experimental data for C_y

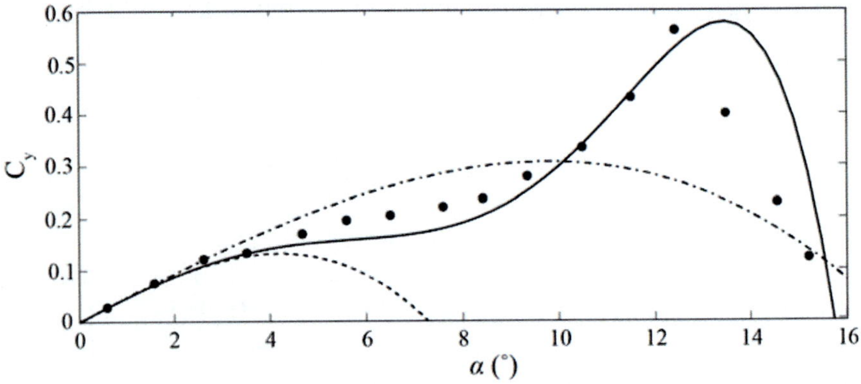

Fig. 5.11 Experimental values for C_y in a square prism (data from Smith, 1962) and three different curve-fitted functions: Continuous line: 7th order fit; Dashed-dotted line: 3rd order fit with $b_3 = -168$, and Dotted line: 3rd order fit with $a_3 = -31$ (Blevins, 2001)

versus angle of attack for the case of square prism placed in flow with its side facing the flow. Three possible curve fitting functions are shown in the figure.

The first curve is a third-order polynomial fit to the experimental data:

$$C_y(\alpha) = a_1 \frac{\dot{y}}{U} + a_3 \left(\frac{\dot{y}}{U}\right)^3, \tag{5.78}$$

where $a_1 = 2.7$ and $a_3 = -31$ (Blevins, 2001). This curve follows the experimental data at small angles very closely, but deviates from them for $\alpha > 4°$.

The second curve is another third-order polynomial approximation, but with different coefficients:

$$C_y(\alpha) = b_1 \frac{\dot{y}}{U} + b_3 \left(\frac{\dot{y}}{U}\right)^3, \tag{5.79}$$

where $b_1 = a_1 = 2.7$ and $b_3 = -168$. This curve represents the data for a wider range of angles of attack, but deviates from the experimental points especially for values close to the peak of C_y at around 13°.

We could also use a fifth-order or a seventh-order polynomial to fit the data. The order of the polynomial is decided by how closely we need to represent the experimental data. The third curve in Fig. 5.11 is a seventh-order polynomial, which gives the closest fit to the experimental data and represents the peaks of the C_y values as well. This seventh-order fit is

$$C_y(\alpha) = b_1 \frac{\dot{y}}{U} + b_3 \left(\frac{\dot{y}}{U}\right)^3 + b_5 \left(\frac{\dot{y}}{U}\right)^5 + b_7 \left(\frac{\dot{y}}{U}\right)^7, \tag{5.80}$$

where $b_1 = 2.7$, $b_3 = -168$, $b_5 = 6270$, and $b_7 = -59900$ (Parkinson & Smith, 1964). If we remove the terms of order 5 and 7 of the seventh-order polynomial fit, we will get the second version of the third-order polynomial fit (Eq. 5.79).

5.1 *The System's Response for the Third-Order Model—Supercritical Instability*

For a first model of galloping of a square cross section, we use the first cubic approximation (Eq. 5.78) for the right-hand side. By using Eq. (5.78) for C_y and plugging it into Eq. (5.77), we find the following nonlinear equation:

$$m\ddot{y} + 2\zeta m\omega_n \dot{y} + ky = \frac{1}{2}\rho h U^2 \left[a_1 \frac{\dot{y}}{U} + a_3 \left(\frac{\dot{y}}{U}\right)^3\right], \tag{5.81}$$

which can be re-ordered as:

$$m\ddot{y} + \left(2\zeta m\omega_n U - \frac{1}{2}\rho h U^2 a_1\right)\frac{\dot{y}}{U} - \frac{1}{2}\rho h U^2 a_3 \left(\frac{\dot{y}}{U}\right)^3 + ky = 0. \tag{5.82}$$

This equation is a third-order nonlinear ODE, with a cubic damping term. This nonlinear damping term is the only new term in comparison with the linear model of the system. By focusing on the two flow-induced damping terms, the linear and the nonlinear ones, we notice that the linear term has a negative coefficient (remember that a_1 is positive) and the nonlinear term has a positive coefficient (a_3 is negative). This means that the nonlinear flow-induced damping term always opposes the linear term. Note that $\dot{y}$ and $\dot{y}^3$ have the same sign. Therefore, when $\dot{y}/U$ is small, $\dot{y}/U > (\dot{y}/U)^3$, and the overall damping of the system is negative (with the assumption of small structural damping). For large values of $\dot{y}/U$, however, $\dot{y}/U < (\dot{y}/U)^3$ and the system will have a positive damping.

If the system parameters are known, we can pick a flow velocity as an input and solve Eq. (5.82) numerically, for a given set of initial conditions. For each flow velocity, the numerical solution will give us a time history of displacement, y. Figures 5.12 and 5.13 show samples of time histories and *phase planes* found after

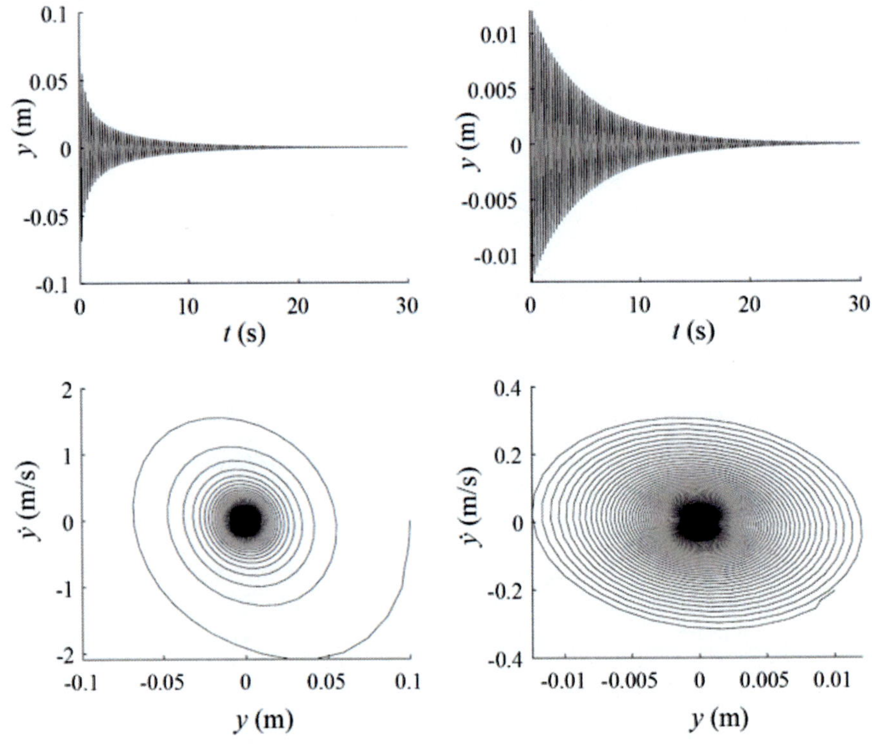

Fig. 5.12 Sample time histories and phase planes for a static response, with two different initial conditions

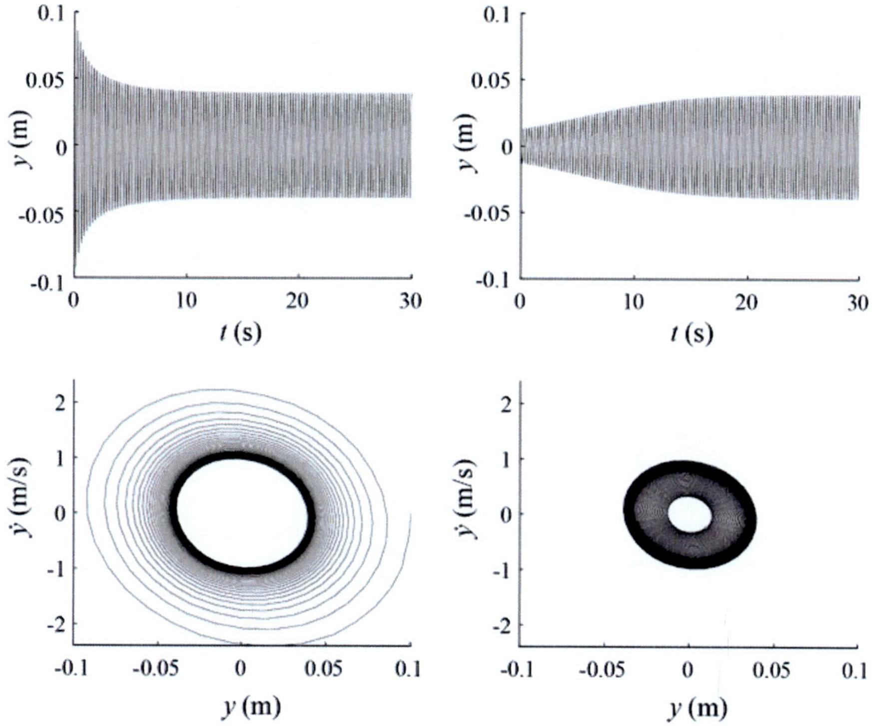

Fig. 5.13 Sample time histories and phase planes for limit cycle oscillations of a galloping square, with two different initial conditions

solving the nonlinear equation for a system with the same parameters as the parameters of the linear system discussed in Sect. 3.8. A phase plane is the plot of velocity, $\dot{y}$, versus displacement, y, and is used extensively in analyzing the response of nonlinear systems. For low flow velocities, the response of the system converges to zero and the system goes back to its original equilibrium state as shown for a sample low flow velocity in Fig. 5.12. Any initial condition at this flow velocity will result in the same solution, as this is the only stable solution of the system. Sample results with different initial conditions are shown in the figure, and they all decay toward zero.

At higher flow velocities, the response of the system converges to an oscillatory motion with a nonzero amplitude and frequency, as shown for a sample flow velocity in Fig. 5.13. As the time histories and phase planes in this plot show, the response of the system is independent from the initial condition. Any initial condition results in the same steady-state response. This steady-state response, which is the same for several different initial conditions, is called a *limit cycle response*. For the sample initial conditions shown in the phase plane plots, the initial condition that is located inside the limit cycle results in a trajectory that grows outward and reaches the limit cycle as the solution converges, and the initial condition that is located outside the limit cycle creates a trajectory that moves inward to approach the limit cycle. We call

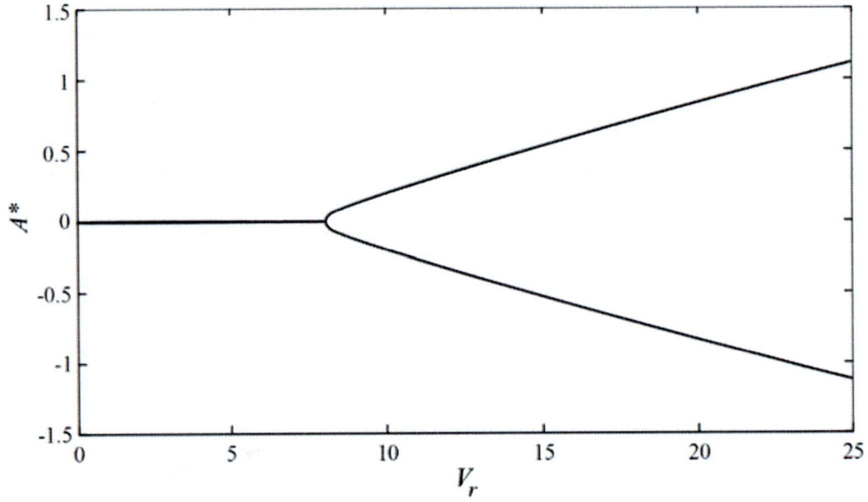

Fig. 5.14 Bifurcation diagram for a 1 DOF square prism in flow, modeled with the 3rd order approximation of Eq. (5.78)

this a limit cycle, because several different initial conditions converge to it, and this seems to be a "limiting" response for a range of initial conditions, as opposed to the response of a linear system, which would converge to a different amplitude for any given initial condition.

The overall behavior of the system can be summarized in a plot of amplitude versus ftlow velocity, in which the amplitude of the steady-state response is plotted for different reduced velocities. Figure 5.14 shows this diagram for the dimensionless amplitude of oscillations of a square prism, $A^* = A/h$, versus the reduced velocity, defined as $V_r = U/f_n h$. At low reduced velocities, the structure stays at its original equilibrium position. Then at a critical reduced velocity of $V_r = 8.3$ in this example, the original stable solution of the system loses its stability, and galloping starts. For reduced velocities larger than this critical value, a limit cycle response is observed. This change of stability in the system response is called a bifurcation. The plot that shows the state of the system for different values of an independent parameter (reduced velocity in this case) is called a *bifurcation diagram*. For reduced velocities higher that the critical, the bifurcation diagram shows the maximum and minimum values of the limit cycle oscillations at each reduced velocity. A complete bifurcation diagram would have shown the unstable solution after the critical point as well (that would be an unstable solution at $A^* = 0$ for values larger than $V_r = 8.3$), but a numerical solution of the ODE (which is what we have used here) cannot converge to an unstable solution. There are ways (numerical and analytical) to obtain the unstable solutions of the system, but we do not discuss them here.

The particular instability that is observed in this case is called a *Hopf bifurcation*. A Hopf bifurcation occurs when a static solution (the original equilibrium position in this case) becomes unstable and a stable limit cycle oscillation emerges. In general, a system can undergo several bifurcations as an independent variable (the reduced velocity in this case) is varied. The type of Hopf bifurcation that is observed in this bifurcation diagram is called a *supercritical* Hopf bifurcation. This means that the single stable branch (the original equilibrium solution) gives way to solutions that are stable, and for values larger than the critical, the amplitude of the response increases smoothly from zero without any sudden jump. Another type of Hopf bifurcation is called a *subcritical* Hopf bifurcation and refers to the cases where the stable solution that is observed after the critical point is at a magnitude much larger than zero, immediately after the critical value, and therefore the system jumps to a higher magnitude right after the critical value. We will observe a subcritical instability later in Sect. 5.2.

Note that the critical value for the onset of oscillations in Fig. 5.14 is $V_r = 8.3$, which is the same value that we found using the linear equation of this system in Sect. 3.8. After the critical point, the original equilibrium position is not stable anymore. This means that although the origin is still a mathematical solution (the trivial solution) of the equation of motion, it cannot be observed physically in an experiment as a long-term solution, because any disturbance will result in converging to a limit cycle oscillation, as suggested by the plots of Fig. 5.13.

5.2 A Seventh-Order Model and Subcritical Instability

Here, we discuss the changes in the solution, if we consider the seventh-order function to fit the experimental data for C_y versus α (Eq. 5.80). A seventh-order polynomial expansion gives a better curve fit into the experimental data, and it is expected to give more reliable results.

A bifurcation diagram that summarizes the results for this case is given in Fig. 5.15. A clear qualitative difference between the results based on the seventh-order fit and those based on the third-order fit is observed for $V_r \sim 10.5$–15. For the results based on the seventh-order fit, the prism is initially stable and becomes unstable at a critical reduced velocity of $V_{r\_cr} = 8.3$, similar to what we observed in the linear analysis and also by using the third-order model. It is expected to get the same critical flow velocity using all these models, because before this instability, the structure does not move, and therefore the small-amplitude assumption that is the main assumption for a linear system is still valid. For reduced velocities larger than the critical, the prism starts to oscillate, as also predicted by the third-order model, but with a smaller magnitude. Then over a range of $V_r \sim 10.5$–15, a second limit cycle region is observed in the bifurcation diagram of Fig. 5.15. This means that in this range, the prism can oscillate with either of these two magnitudes, depending on the initial conditions. After this range, for $V_r > 15$, only one solution exists, similar to what we observed in the third-order bifurcation diagram, but with a higher magnitude compared with that case, and the prism behaves very similarly in both cases for the rest of the V_r range.

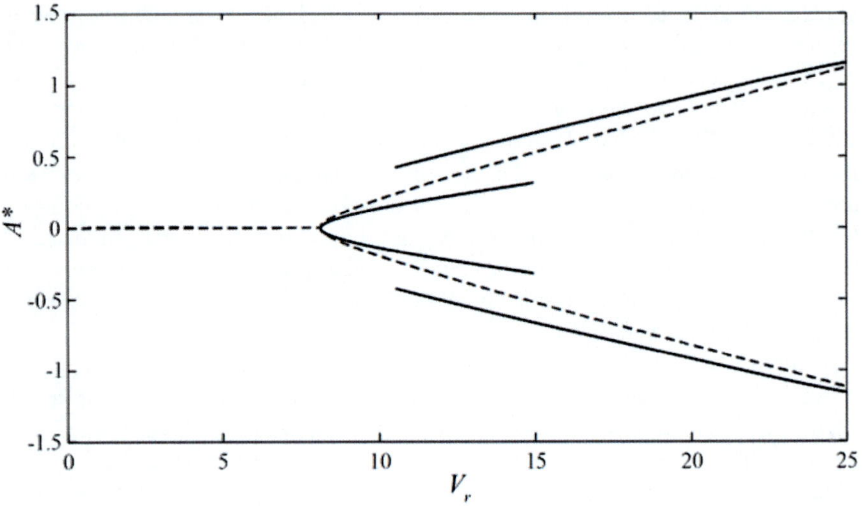

Fig. 5.15 Bifurcation diagram for results based on the 3rd order (dashed line) and the 7th order (continuous line) models

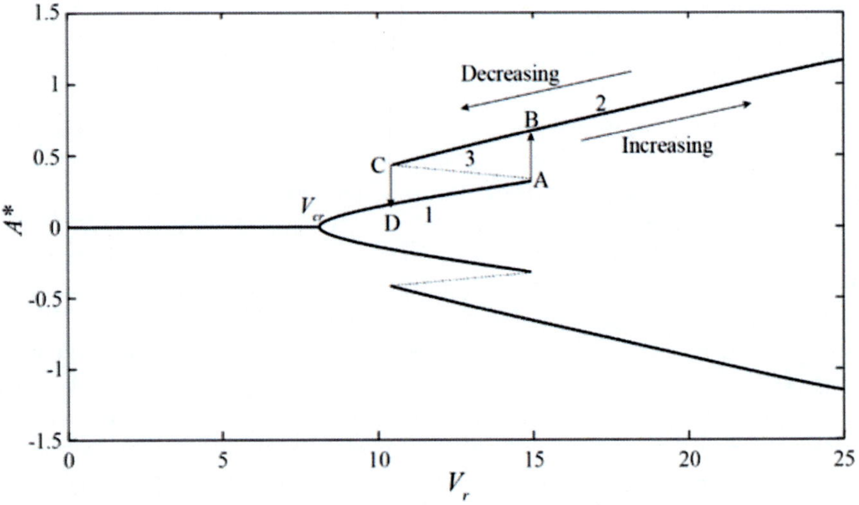

Fig. 5.16 Bifurcation diagram for results based on the 7th order model

If we focus on the response based on the seventh-order fit, we observe that after the first point of instability at V_{r_cr}, the prism undergoes limit cycle oscillations with amplitudes that are shown by branch 1 in the bifurcation diagram (Fig. 5.16). As the reduced velocity is increased, at $V_r \sim 15$, this branch ends, which implies that another bifurcation has occurred at point A. At slightly higher values of V_r, the amplitude of limit cycle oscillations jumps into the values on branch 2 (point B), and it follows that branch all the way to the maximum reduced velocity shown here. Then as the reduced velocity is decreased, the magnitude stays on branch 2 all the way down to $V_r \sim 10$ (point C). If we decrease the reduced velocity even further, then the magnitude drops suddenly to branch 1, at point D. Therefore, in the ranges of D-A and C-B, there are two possible solutions. The bifurcation at point A is a subcritical bifurcation, since there is a sudden jump from one stable response to a new stable response at values slightly larger than the critical. After this jump, the amplitude of oscillations follows a continuous trend (as long as there is no other instability). In this case, branch 1 does not really disappear. It becomes unstable, and moves backward (to the left, branch 3). Branch 3 cannot be observed experimentally (or by solving the ODE numerically), because it represents an unstable limit cycle response, but it can be obtained mathematically. This branch goes backward up to another critical point (point C) at which another bifurcation occurs, and the resulting branch (branch 2) bends again and moves to the right and represents a stable limit cycle oscillation. The bifurcation that occurs at point C is called a saddle-node coalescence of limit cycles. Saddle-node is a bifurcation at which two solutions (one stable and one unstable) emerge from no solution! If we look at an area very close to point C and to its left, there is no solution, and as V_r is increased, two solutions emerge from nothing, and since these two solutions are limit cycles, the bifurcation is called a saddle-node coalescence of limit cycles. Strogatz (2015) discusses different types of bifurcations that could be observed in a nonlinear system.

Chapter 6
Vibrations of Continuous Structures

1 Introduction

Starting in this chapter, we will focus on continuous structures and study their interaction with flow. A continuous structure such as a beam or a string is a structure in which the response is a function of time and space. If we consider a string (e.g., a guitar string) as it oscillates, different points on the string move differently. The motion at each point is a function of time, but it is not necessarily the same function for different points along the length of the string. Therefore, the response of a string is a function of time and space. The equations that represent such systems are called partial differential equations (PDEs). PDEs include derivatives with respect to both time and space.

Before we study how continuous structures interact with flow, we focus on continuous structures with no flow around them, similar to what we did in Chap. 3 for 1 DOF and 2 DOF structures before we added any flow to the system. In this chapter, we will study beams and strings, since they are the continuous systems that are widely used in FSI problems. Shells and plates are also continuous structures, but we do not consider them in this chapter.

2 Linear Equation of Motion for a Beam under External Tension

Assume a beam with length L, density ρ, and cross-sectional area A, which can oscillate in the direction perpendicular to the long axis of the beam (y-direction in Fig. 6.1a). We assume a continuous external force, $f(x,t)$ per unit length, is applied on the beam as well and that the beam is under an external tension, T, that is applied along its length. To derive the equation of motion that governs this system, we consider the free body diagram of an arbitrary element of the beam with an

Y. Modarres-Sadeghi, *Introduction to Fluid-Structure Interactions*,
https://doi.org/10.1007/978-3-030-85884-1_6

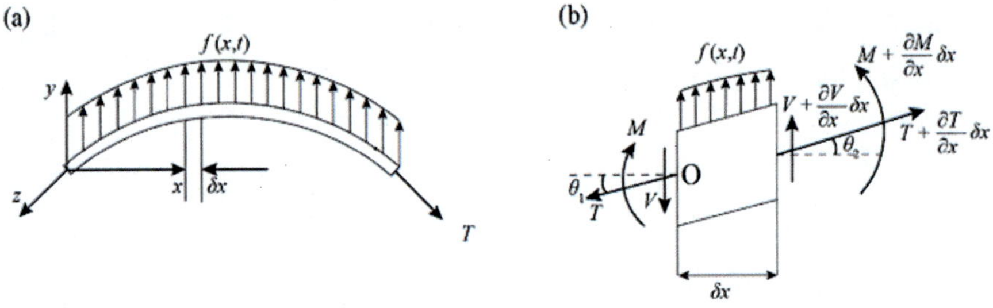

Fig. 6.1 (**a**) A schematic of a beam that is free to oscillate in the y-direction and (**b**) the free body diagram of a beam element at an arbitrary point along its length, at a distance x from the origin

infinitesimal length, δx, located at a point with a distance x from one end of the beam (the left end in this case). We consider this end of the beam, the origin. The free-body diagram (Fig. 6.1b) shows the tension force, the shear force, the distributed external force, and the moments that act on this element. We call the shear force on the left-hand side of the element, V. This force is slightly different on the right-hand side of the element: $V + (\partial V/\partial x)\delta x$, where $(\partial V/\partial x)\delta x$ represents the changes in V over the small element of δx. Similarly, the moments on the left-hand side and the right-hand side are slightly different: M and $M + (\partial M/\partial x)\delta x$, respectively. The slopes of the beam at the two ends of the element are different too: The angle that the beam makes with the horizontal direction (its original equilibrium position) is θ_1 on the left-hand side of the element, and θ_2 on the right-hand side. These are also the angles that the tension force makes with the horizontal direction at those points. We assume these angles are small, since our goal is to derive the linear equation of motion. With the small-angle assumption, we have:

$$\sin\theta_1 \simeq \tan\theta_1 = \left.\frac{\partial y}{\partial x}\right|_{x_1}, \tag{6.1}$$

$$\sin\theta_2 \simeq \tan\theta_2 = \left.\frac{\partial y}{\partial x}\right|_{x_2}, \tag{6.2}$$

and

$$\cos\theta_1 = \cos\theta_2 \simeq 1. \tag{6.3}$$

Using Newton's second law in the x-direction, we find:

$$\sum F_x = 0 \quad \Rightarrow T\cos\theta_1 = (T + \frac{\partial T}{\partial x})\cos\theta_2, \tag{6.4}$$

which, since $\cos\theta_1 = \cos\theta_2 \simeq 1$, results in:

$$\frac{\partial T}{\partial x} = 0. \tag{6.5}$$

which means that the tension stays constant in the x-direction. Newton's second law in the y-direction gives:

$$\sum F_y = ma_y$$
$$\Rightarrow -V + \left(V + \frac{\partial V}{\partial x}\delta x\right) + f\delta x - T\frac{\partial y}{\partial x}\bigg|_{x_1} + T\frac{\partial y}{\partial x}\bigg|_{x_2} = \rho A\delta x\frac{\partial^2 y}{\partial t^2}, \tag{6.6}$$

or

$$\frac{\partial V}{\partial x}\delta x + f\delta x - T\frac{\partial y}{\partial x}\bigg|_{x_1} + T\frac{\partial y}{\partial x}\bigg|_{x_2} = \rho A\delta x\frac{\partial^2 y}{\partial t^2}. \tag{6.7}$$

Every term in Eq. (6.7), except the two terms with T, includes δx. If we write these two terms with T in such a way that a δx also shows up in them, then we can drop δx from all terms. In that case, the equation becomes independent from the size and location of the element that we picked for our free body diagram and it will be valid for any point along the length of the beam. Note that in the two terms with T, the slope is calculated at two different points: x_1 and x_2. To relate one to the other, we use the Taylor expansion to expand $T(\partial y/\partial x)|_{x_2}$ about $T(\partial y/\partial x)|_{x_1}$:

$$T\frac{\partial y}{\partial x}\bigg|_{x_2} = T\frac{\partial y}{\partial x}\bigg|_{x_1} + \frac{\partial}{\partial x}\left(T\frac{\partial y}{\partial x}\right)\delta x. \tag{6.8}$$

Since we are deriving a linear equation, the Taylor expansion does not include any higher-order terms. Then by plugging Eq. (6.8) into Eq. (6.7), we get:

$$\frac{\partial V}{\partial x}\delta x + f\delta x - T\frac{\partial y}{\partial x}\bigg|_{x_1} + T\frac{\partial y}{\partial x}\bigg|_{x_1} + \frac{\partial}{\partial x}\left(T\frac{\partial y}{\partial x}\right)\delta x = \rho A\delta x\frac{\partial^2 y}{\partial t^2}, \tag{6.9}$$

which can be simplified to:

$$\frac{\partial V}{\partial x}\delta x + f\delta x + \frac{\partial}{\partial x}\left(T\frac{\partial y}{\partial x}\right)\delta x = \rho A\delta x\frac{\partial^2 y}{\partial t^2}, \tag{6.10}$$

and now we can drop δx from all terms to get:

$$\frac{\partial V}{\partial x} + f + \frac{\partial}{\partial x}\left(T\frac{\partial y}{\partial x}\right) = \rho A\frac{\partial^2 y}{\partial t^2}. \tag{6.11}$$

And since T is assumed to be constant:

$$\frac{\partial V}{\partial x}+f+T\frac{\partial^2 y}{\partial x^2}=\rho A\frac{\partial^2 y}{\partial t^2}. \tag{6.12}$$

In this equation, there are two unknowns: y and V, and therefore we need another equation to relate the two. To find the second equation, we apply the torque version of Newton's second law to the free body diagram of Fig. 6.1b and write:

$$\begin{aligned}&\sum M_O=0,\\&-M+\left(M+\frac{\partial M}{\partial x}\delta x\right)+\left(V+\frac{\partial V}{\partial x}\delta x\right)\delta x-[f(x,t)\delta x]\frac{\delta x}{2}=0,\end{aligned} \tag{6.13}$$

or

$$\frac{\partial M}{\partial x}\delta x+V\delta x+\frac{\partial V}{\partial x}(\delta x)^2-\frac{1}{2}f(x,t)(\delta x)^2=0. \tag{6.14}$$

With the assumption that δx is small, then $(\delta x)^2\simeq 0$. In fact, when we assumed δx is small, we assumed that it is small enough such that $(\delta x)^2\simeq 0$. Then Eq. (6.14) becomes:

$$\frac{\partial M}{\partial x}\delta x+V\delta x=0, \tag{6.15}$$

or

$$\frac{\partial M}{\partial x}+V=0. \tag{6.16}$$

This is an equation that relates the moment to the shear force, and since the shear force also appeared in Eq. (6.12), by plugging Eq. (6.16) into Eq. (6.12), we find:

$$\frac{\partial}{\partial x}\left(-\frac{\partial M}{\partial x}\right)+f+T\frac{\partial^2 y}{\partial x^2}=\rho A\frac{\partial^2 y}{\partial t^2}, \tag{6.17}$$

or:

$$-\frac{\partial^2 M}{\partial x^2}+f+T\frac{\partial^2 y}{\partial x^2}=\rho A\frac{\partial^2 y}{\partial t^2}. \tag{6.18}$$

In this equation, y and M are the unknowns—in fact we have replaced one unknown, V, by another, M. However, we can relate M to y, based on what we remember from Strength of Materials: for a deflected beam, one can relate the bending moment to the second derivative of displacement with respect to space as:

$$M = EI\frac{\partial^2 y}{\partial x^2}, \tag{6.19}$$

where E is the Young's modulus of elasticity and I is the cross-sectional area moment of inertia. By plugging Eq. (6.19) into Eq. (6.18), we find:

$$-\frac{\partial^2}{\partial x^2}\left(EI\frac{\partial^2 y}{\partial x^2}\right) + f + T\frac{\partial^2 y}{\partial x^2} = \rho A\frac{\partial^2 y}{\partial t^2}, \tag{6.20}$$

which can be re-written as:

$$\rho A\frac{\partial^2 y}{\partial t^2} + \frac{\partial^2}{\partial x^2}\left(EI\frac{\partial^2 y}{\partial x^2}\right) - T\frac{\partial^2 y}{\partial x^2} = f. \tag{6.21}$$

For a beam with constant cross-sectional area and same material (constant material properties) throughout its length, EI is constant and the equation of motion becomes:

$$\rho A\frac{\partial^2 y}{\partial t^2} + EI\frac{\partial^4 y}{\partial x^4} - T\frac{\partial^2 y}{\partial x^2} = f. \tag{6.22}$$

This is the linear equation of motion for a beam with a constant cross section and of the same material throughout its length, under a continuous external force, f, and a constant external tension, T. The first term in the equation is an inertia term, the second and the third terms are stiffness terms, and there is no damping term, because we have assumed that the beam has no structural damping. If we assume nonzero structural damping, then the equation of motion becomes:

$$\rho A\frac{\partial^2 y}{\partial t^2} + c\frac{\partial y}{\partial t} + EI\frac{\partial^4 y}{\partial x^4} - T\frac{\partial^2 y}{\partial x^2} = f, \tag{6.23}$$

where c is a damping coefficient.

Equation (6.23) consists of an inertial term, a damping term, and a stiffness term, similar to the form of the equation of motion of a 1 DOF mass-spring-damper that we discussed in Chap. 3.

If no external tension is applied, and in the case of no structural damping, the equation becomes,

$$\rho A\frac{\partial^2 y}{\partial t^2} + EI\frac{\partial^4 y}{\partial x^4} = f, \tag{6.24}$$

which is the simplest form of a beam equation. However, if we assume that the term with EI is negligible compared with the term with T, and there is no structural damping, then Eq. (6.23) becomes:

$$\rho A \frac{\partial^2 y}{\partial t^2} - T \frac{\partial^2 y}{\partial x^2} = f. \tag{6.25}$$

One case where we can make this assumption is a guitar string, where the motion is dominated by the tension that is applied to the structure. Equation (6.25) is known as the equation of motion of a tension-dominated beam (i.e., the stiffness is dominated by tension, and not by the flexural rigidity). This equation is also known as the equation of motion for a string. A simple way to distinguish between a string and a beam is that a string falls if it is fixed at one end and free at the other end with no applied tension. A beam, on the other hand, can stay mainly horizontal if it is clamped at one end but free at its other end, even if there is no external tension. Later, in Sect. 6, we will discuss a quantitative method that we can use to distinguish the dominance of tension or flexural rigidity terms.

3 The Natural Frequencies and Mode Shapes of a Beam

The first step toward understanding the dynamic response of a beam is to calculate its natural frequencies. As we will see soon, a beam has infinite natural frequencies. We can think of the response of a beam as a combination of responses of infinite flexibly mounted point masses that are connected to each other. This clearly constitutes a system with infinite degrees of freedom. Here, however, we think of the response of a beam as a combination of responses of several beam modes. In what follows, we will discuss what is meant by modes of a beam, and we will see that every beam has infinite modes, and infinite natural frequencies.

In order to find the natural frequencies and mode shapes of a beam with no external tension, we use Eq. (6.24) with no external force, i.e., $f = 0$. We remember from the discussion on the natural frequency of a 1 DOF system in Chap. 3 that the natural frequency does not depend on the external forces that act on the system, neither does it depend on the structural damping. Therefore, the equation of motion that will give us the natural frequencies of a beam reads as:

$$\frac{\partial^2 y}{\partial t^2} + \frac{EI}{\rho A} \frac{\partial^4 y}{\partial x^4} = 0. \tag{6.26}$$

The coefficient of the second term is a contact for a defined system and depends on the properties of the beam. We use a constant, c_b, for this coefficient:

$$c_b = \sqrt{\frac{EI}{\rho A}}. \tag{6.27}$$

Then, the equation of motion can be written as

$$\frac{\partial^2 y}{\partial t^2} + c_b^2 \frac{\partial^4 y}{\partial x^4} = 0. \tag{6.28}$$

This is a linear partial differential equation of order 2 in time and order 4 in space. To solve it, we need 2 initial conditions and 4 boundary conditions.

The initial conditions should be the zeroth and the first derivatives of the unknown, y, with respect to time, i.e., displacement and velocity of the beam. Since y is a function of time and space, $y(x,t)$, the initial conditions are written as:

$$y(x,0) = y_0(x), \tag{6.29}$$

and

$$\frac{\partial y}{\partial t}(x,0) = v_0(x), \tag{6.30}$$

in which $y_0(x)$ and $v_0(x)$ are, respectively, displacement and velocity of the beam at time zero.

Of the four boundary conditions that are needed for this PDE, each end of the beam contributes two. These boundary conditions could be the beam's deflection: y, its slope: $\partial y/\partial x$, its bending moment, $EI\partial^2 y/\partial x^2$, or its shear force: $EI\partial^3 y/\partial x^3$ at either of the two ends of the beam (two for each end).

Once the boundary conditions are set, in order to solve the PDE, we use the method of separation of variables. This method assumes that the solution, $y(x,t)$, can be written as multiplication of two functions: a function of x and a function of t, i.e.,

$$y(x,t) = X(x)T(t). \tag{6.31}$$

After making this assumption, we can plug $y(x,t)$ into the original PDE (Eq. 6.28) and get:

$$X(x)\ddot{T}(t) + c_b^2 X''''(x)T(t) = 0, \tag{6.32}$$

which can be re-arranged into

$$c_b^2 \frac{X''''(x)}{X(x)} = -\frac{\ddot{T}(t)}{T(t)}, \tag{6.33}$$

where dot is derivative with respect to time and prime is derivative with respect to space.

In Eq. (6.33), we notice that the left-hand side is a function of x and the right-hand side is a function of t. The only way that these two functions could be equal is if they both are equal to a constant, which we assume to be ω^2:

$$c_b^2 \frac{X''''(x)}{X(x)} = -\frac{\ddot{T}(t)}{T(t)} = \omega^2. \tag{6.34}$$

This gives us two equations:

$$-\frac{\ddot{T}(t)}{T(t)} = \omega^2, \tag{6.35}$$

and

$$c_b^2 \frac{X''''(x)}{X(x)} = \omega^2. \tag{6.36}$$

The first equation (Eq. 6.35) is a second-order ODE, and is the same equation that we discussed in Chap. 3, when we considered a 1 DOF mass-spring system with no damping. By solving that equation, we found (and in fact defined) the natural frequency of a 1 DOF system. Here, again, we can use this equation to find the natural frequencies of the system. We will soon see how we will get more than one natural frequency (and therefore we used plural in the previous sentence). The first equation:

$$\ddot{T}(t) + \omega^2 T(t) = 0, \tag{6.37}$$

has a general solution of:

$$T(t) = A \sin \omega t + B \cos \omega t, \tag{6.38}$$

where A and B are constants, which depend on the initial conditions. Before we find A and B, we consider the second equation (Eq. 6.36), which can be written as:

$$X''''(x) - \frac{\omega^2}{c_b^2} X(x) = 0, \tag{6.39}$$

or

$$X''''(x) - \beta^4 X(x) = 0, \tag{6.40}$$

where

Table 6.1 Typically observed boundary conditions in beams

Type of boundary	Boundary conditions	Schematic
Free	No bending moment: $EIy^{''} = 0$ No shear force: $EIy^{'''} = 0$	
Simply supported (hinged)	No displacement: $y = 0$ No bending moment: $EIy^{''} = 0$	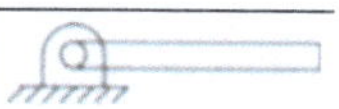
Clamped	No displacement: $y = 0$ No slope: $y' = 0$	

$$\beta^4 \equiv \frac{\omega^2}{c_b^2}. \tag{6.41}$$

This equation is a fourth-order boundary value problem, and to solve it, we consider a general solution of

$$X(x) = a_1 \sin \beta x + a_2 \cos \beta x + a_3 \sinh \beta x + a_4 \cosh \beta x, \tag{6.42}$$

in which a_1 to a_4 are constants that we need to find. Besides these 4 unknowns, β is an unknown, because we have not yet found the natural frequencies of the system. To find the constants a_1 to a_4 in this general solution, we need to decide on the boundary conditions. Table 6.1 gives a summary of the boundary conditions that are usually observed in beams. In the following section, we will consider a cantilevered beam as an example to see how the natural frequencies and mode shapes of a beam are derived.

4 Example: Deriving the Natural Frequencies and Mode Shapes of a Cantilevered Beam

Here, we assume that the beam is a cantilevered beam: Clamped at its left end and free at its right end. The boundary conditions for a cantilevered beam are zero displacement and slope at $x = 0$, and zero bending moment and shear force at $x = L$:

$$\begin{aligned} y(0, t) &= 0, \\ y'(0, t) &= 0, \\ EIy''(L, t) &= 0, \\ EIy'''(L, t) &= 0. \end{aligned} \tag{6.43}$$

Following the assumption of $y(x,t) = X(x)T(t)$, these boundary conditions can be written in terms of X as:

$$\begin{aligned} y(0,t) &= 0 \Rightarrow X(0) = 0, \\ y'(0,t) &= 0 \Rightarrow X'(0) = 0, \\ EIy''(L,t) &= 0 \Rightarrow X''(L) = 0, \\ EIy'''(L,t) &= 0 \Rightarrow X'''(L) = 0. \end{aligned} \tag{6.44}$$

Using the general solution (Eq. 6.42) to satisfy these boundary conditions, we will get the following equations:

$$X(0) = 0 \Rightarrow a_2 + a_4 = 0, \tag{6.45}$$

$$X'(0) = 0 \Rightarrow a_1\beta + a_3\beta = 0, \tag{6.46}$$

$$X''(L) = 0 \Rightarrow -a_1\beta^2 \sin\beta L - a_2\beta^2 \cos\beta L + a_3\beta^2 \sinh\beta L + a_4\beta^2 \cosh\beta L = 0, \tag{6.47}$$

$$X'''(L) = 0 \Rightarrow -a_1\beta^3 \cos\beta L + a_2\beta^3 \sin\beta L + a_3\beta^3 \cosh\beta L + a_4\beta^3 \sinh\beta L = 0. \tag{6.48}$$

We can drop the β's from all terms and write the equations in the matrix form:

$$\begin{bmatrix} 0 & 1 & 0 & 1 \\ 1 & 0 & 1 & 0 \\ -\sin\beta L & -\cos\beta L & \sinh\beta L & \cosh\beta L \\ -\cos\beta L & \sin\beta L & \cosh\beta L & \sinh\beta L \end{bmatrix} \begin{Bmatrix} a_1 \\ a_2 \\ a_3 \\ a_4 \end{Bmatrix} = \begin{Bmatrix} 0 \\ 0 \\ 0 \\ 0 \end{Bmatrix}, \tag{6.49}$$

or

$$\mathbf{AV} = \mathbf{0}, \tag{6.50}$$

where

$$\mathbf{A} = \begin{bmatrix} 0 & 1 & 0 & 1 \\ 1 & 0 & 1 & 0 \\ -\sin\beta L & -\cos\beta L & \sinh\beta L & \cosh\beta L \\ -\cos\beta L & \sin\beta L & \cosh\beta L & \sinh\beta L \end{bmatrix}, \tag{6.51}$$

and

$$\mathbf{V} = \begin{Bmatrix} a_1 \\ a_2 \\ a_3 \\ a_4 \end{Bmatrix}. \tag{6.52}$$

Note that in these four equations, we have five unknowns: a_1, a_2, a_3, a_4, and β. One easy way to satisfy Eq. (6.50) is to assume $\mathbf{V} = \mathbf{0}$, which means that all coefficients are equal to zero: $a_1 = a_2 = a_3 = a_4 = 0$. This means that $X(x) = 0$ and therefore $y(x,t) = 0$, which is the *trivial* solution of the system: Any beam with no external force and no disturbance will stay at its original equilibrium position forever. We, however, are looking for cases where the beam oscillates, and therefore we are looking for nontrivial solutions. For a nontrivial solution, the determinant of matrix **A** must be zero (so that the matrix will not have an inverse, and zero cannot be an answer for vector **V**):

$$\det(\mathbf{A}) = 0, \tag{6.53}$$

which results in the following equation:

$$\cos\beta L \cosh\beta L = -1 \tag{6.54}$$

Equation (6.54) is called the *characteristic equation*, and has infinite solutions, which can be found to be:

$$\begin{aligned} \beta_1 L &= 1.875, \\ \beta_2 L &= 4.694, \\ \beta_3 L &= 7.855, \\ \beta_4 L &= 10.996, \\ \beta_5 L &= 14.137, \\ \beta_n L &= \frac{(2n-1)\pi}{2}, \text{ for } n > 5. \end{aligned} \tag{6.55}$$

Since we get infinite solutions for β, we have infinite natural frequencies in a continuous system. We should remember that β and ω (the natural frequency of the system) are related (from Eq. 6.41):

$$\beta_n^4 = \frac{\omega_n^2}{c_b^2} = \frac{\omega_n^2 \rho A}{EI}. \tag{6.56}$$

We use subscript n, where n goes from 1 to infinity, to remind ourselves that we have infinite natural frequencies. Then the natural frequencies can be written in terms of β and the system parameters as

$$\omega_n = \beta_n^2 \sqrt{\frac{EI}{\rho A}}. \tag{6.57}$$

Since βL is the solution of eq. (6.54) (not β itself), we write the natural frequencies in terms of βL:

$$\omega_n = \frac{\beta_n^2 L^2}{L^2} \sqrt{\frac{EI}{\rho A}}. \tag{6.58}$$

Now that we have the natural frequencies of the beam, we can go back to the four equations that satisfied the boundary conditions (Eqs. 6.45–6.48) and use them to relate three of the coefficients to the fourth one in the general solution. To do so, from the first equation (Eq. 6.45) we will have:

$$a_2 = -a_4, \tag{6.59}$$

and from the second equation (Eq. 6.46):

$$a_1 = -a_3. \tag{6.60}$$

Plugging these two relations into the 3rd and the 4th equations (Eqs. 6.47 and 6.48), we will get:

$$\begin{aligned} a_3 \sin \beta L + a_4 \cos \beta L + a_3 \sinh \beta L + a_4 \cosh \beta L = 0, \\ a_3 \cos \beta L - a_4 \sin \beta L + a_3 \cosh \beta L + a_4 \sinh \beta L = 0, \end{aligned} \tag{6.61}$$

and from the first of these two equations:

$$a_3 = -\frac{\cos \beta L + \cosh \beta L}{\sin \beta L + \sinh \beta L} a_4, \tag{6.62}$$

which also gives:

$$a_1 = \frac{\cos \beta L + \cosh \beta L}{\sin \beta L + \sinh \beta L} a_4. \tag{6.63}$$

Now, we have a_1, a_2, and a_3 in terms of a_4. If we plug these into the original form of $X(x)$ (Eq. 6.42):

$$X(x) = a_1 \sin \beta x + a_2 \cos \beta x + a_3 \sinh \beta x + a_4 \cosh \beta x, \tag{6.64}$$

we will get:

$$X(x) = \frac{\cos\beta L + \cosh\beta L}{\sin\beta L + \sinh\beta L} a_4 \sin\beta x - a_4 \cos\beta x - \frac{\cos\beta L + \cosh\beta L}{\sin\beta L + \sinh\beta L} a_4 \sinh\beta x + a_4 \cosh\beta x, \quad (6.65)$$

or

$$X(x) = \left(\cosh\beta x - \cos\beta x - \frac{\cos\beta L + \cosh\beta L}{\sin\beta L + \sinh\beta L} (\sinh\beta x - \sin\beta x) \right) a_4. \quad (6.66)$$

Since a_4 is an arbitrary magnitude, we can drop it, and call the remainder of the function, the *beam's mode shape*. Since X is a function of β and we have infinite β's, we will have infinite X's, and we refer to them as X_n:

$$X_n(x) = \cosh\beta_n x - \cos\beta_n x - \sigma_n(\sinh\beta_n x - \sin\beta_n x), \quad (6.67)$$

where

$$\sigma_n = \frac{\cos\beta_n L + \cosh\beta_n L}{\sin\beta_n L + \sinh\beta_n L}. \quad (6.68)$$

This equation gives mode shapes of a cantilevered beam. The first four modes are shown in Fig. 6.2. In this figure, each mode is normalized such that its maximum is 1. Since we have dropped a coefficient (a_4) from the relation that gives $X_n(x)$, the magnitudes of these modes are not relevant. It is only their shapes that are of interest.

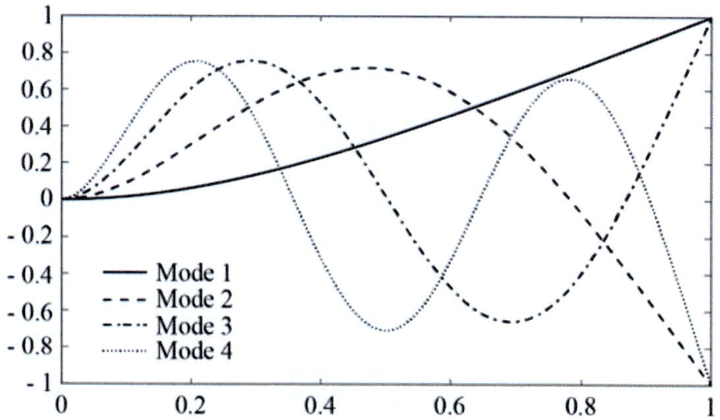

Fig. 6.2 The first four mode shapes of a cantilevered beam

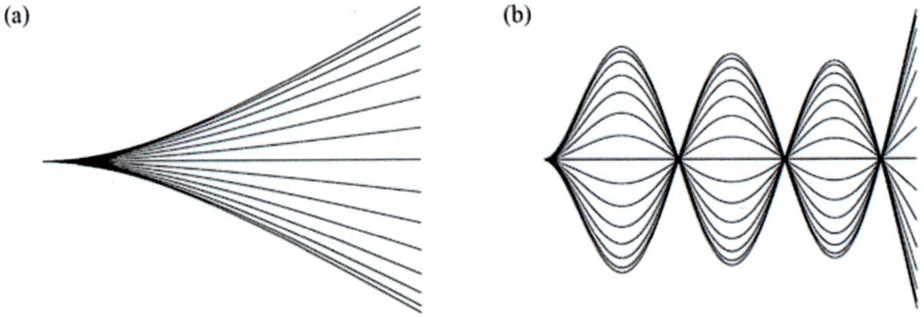

Fig. 6.3 Snapshots of a beam oscillating at its (**a**) 1st mode and (**b**) 4th mode

Each mode shape represents the shape of the beam when it oscillates at the natural frequency that corresponds to that mode. If a cantilevered beam is excited at its base with an external force with a frequency equal to ω_1, then the beam oscillates with a shape that resembles its first mode shape (Fig. 6.3a). If the external force has the frequency of, let's say, the fourth natural frequency, then the beam's oscillations look like its fourth mode shape (Fig. 6.3b). As the beam oscillates in its first mode only, no node is observed along its length (excluding the clamped end) (Fig. 6.3a). A node is a point with zero amplitude of oscillations. As the beam oscillates in its fourth mode, three nodes are observed along its length (Fig. 6.3b). In general, for a cantilevered beam, the number of nodes observed in the mode shape m is $n = m - 1$.

We can find the natural frequencies and mode shapes of beams with other boundary conditions following the same steps, starting from the general form of the mode shapes and by finding the unknown coefficients that satisfy their corresponding boundary conditions. Once the boundary conditions are known, the natural frequencies and mode shapes of the beam will not change. These natural frequencies and mode shapes are given in Table 6.2 for some commonly observed boundary conditions.

5 Notes on Mode Shapes and the Final Form of the Beam's Solution

One very useful feature of the beam mode shapes is that they are *orthogonal* to each other. Mathematically two functions $X_n(x)$ and $X_m(x)$ are orthogonal if:

$$\int_0^L X_m(x)X_n(x)dx = \begin{cases} \frac{L}{2} & \text{for } m = n \\ 0 & \text{for } m \neq n \end{cases} \tag{6.69}$$

Table 6.2 Natural frequencies and mode shapes for beams with various boundary conditions

Boundary conditions	Characteristic equation	Mode shapes	σ_n	$\beta_n L$
Pinned-Pinned (Simply Supported) 	$\sin \beta_n L = 0$	$X_n(x) = \sin \beta_n x$	–	$\beta_n L = n\pi$
Cantilevered 	$\cos \beta_n L \cosh \beta_n L = -1$	$X_n(x) = \cosh \beta_n x - \cos \beta_n x - \sigma_n(\sinh \beta_n x - \sin \beta_n x)$	$\sigma_n = \frac{\cos \beta_n L + \cosh \beta_n L}{\sin \beta_n L + \sinh \beta_n L}$	$\beta_1 L = 1.875$ $\beta_2 L = 4.694$ $\beta_3 L = 7.855$ $\beta_4 L = 10.996$
Free-Free 	$\cos \beta_n L \cosh \beta_n L = 1$	$X_n(x) = \sin \beta_n x + \sinh \beta_n x + \sigma_n(\cos \beta_n x + \cosh \beta_n x)$	$\sigma_n = \frac{\sin \beta_n L - \sinh \beta_n L}{\cosh \beta_n L - \cos \beta_n L}$	$\beta_1 L = 4.730$ $\beta_2 L = 7.853$ $\beta_3 L = 10.996$ $\beta_4 L = 14.137$
Clamped-Clamped 	$\cos \beta_n L \cosh \beta_n L = 1$	$X_n(x) = \sinh \beta_n x - \sin \beta_n x + \sigma_n(\cosh \beta_n x - \cos \beta_n x)$	$\sigma_n = \frac{\sinh \beta_n L - \sin \beta_n L}{\cos \beta_n L - \cosh \beta_n L}$	$\beta_1 L = 4.730$ $\beta_2 L = 7.853$ $\beta_3 L = 10.996$ $\beta_4 L = 14.137$
Clamped-Pinned 	$\tan \beta_n L - \tanh \beta_n L = 0$	$X_n(x) = \sin \beta_n x - \sinh \beta_n x + \sigma_n(\cosh \beta_n x - \cos \beta_n x)$	$\sigma_n = \frac{\sin \beta_n L - \sinh \beta_n L}{\cos \beta_n L - \cosh \beta_n L}$	$\beta_1 L = 3.927$ $\beta_2 L = 7.069$ $\beta_3 L = 10.210$ $\beta_4 L = 13.352$
Pinned-Free 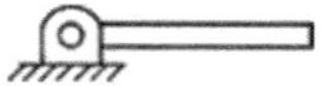	$\tan \beta_n L - \tanh \beta_n L = 0$	$X_n(x) = \sin \beta_n x + \sigma_n \sinh \beta_n x$	$\sigma_n = \frac{\sin \beta_n L}{\sinh \beta_n L}$	$\beta_1 L = 3.927$ $\beta_2 L = 7.069$ $\beta_3 L = 10.210$ $\beta_4 L = 13.352$

The orthogonality of mode shapes means that each mode shape of the beam has no information about the other mode shapes of the beam. This is very similar to the coordinates of a three-dimensional space, x, y, and z. These are three "orthogonal" directions. To know where a point in space is located, we need to know its coordinates in all three directions, because information in one direction does not tell us anything about the point's location in the other two directions. This is the same for the modes of a beam: since they are orthogonal, if we know the response of mode m, that does not tell us anything about the response of mode n. Then by using orthogonal mode shapes, we represent the response of the system in an N-dimensional space, the bases of which are the N orthogonal mode shapes of the beam.

After finding $T(t)$ and $X(x)$, we can go back to the original assumption of

$$y(x,t) = X(x)T(t), \tag{6.70}$$

and find the final solution of the equation of motion, which will have the form:

$$y(x,t) = X(x)(A \sin \omega t + B \cos \omega t). \tag{6.71}$$

Since the beam has infinite modes, Eq. (6.71) should be written as:

$$y(x,t) = \sum_{n=1}^{\infty} X_n(x)(A_n \sin \omega_n t + B_n \cos \omega_n t), \tag{6.72}$$

to include the contributions of all modes to the beam's response.

We need to use the initial conditions to find A_n and B_n. For initial conditions we had:

$$y(x,0) = y_0(x), \tag{6.73}$$

and

$$\dot{y}(x,0) = v_0(x), \tag{6.74}$$

which we plug into the general solution for $y(x,t)$ (Eq. 6.72), to find:

$$\sum_{n=1}^{\infty} X_n(x)(B_n) = y_0(x) \tag{6.75}$$

and

$$\sum_{n=1}^{\infty} X_n(x)(\omega_n A_n) = v_0(x). \tag{6.76}$$

Then, we use the orthogonality of mode shapes to find A_n and B_n. To do so, we multiply both sides of Eq. (6.75) by X_m and integrate the resulting equation from 0 to L:

$$\sum_{n=1}^{\infty} \int_0^L X_m(x)X_n(x)(B_n)dx = \int_0^L X_m(x)y_0(x)dx. \tag{6.77}$$

Because of the orthogonality of the mode shapes (Eq. 6.69), all the terms on the left-hand side of eq. (6.77) are zero except for one term, where $n = m$, which is equal to $L/2$. Therefore,

$$\frac{L}{2}B_m = \int_0^L X_m(x)y_0(x)dx, \tag{6.78}$$

which can be re-written with subscript n instead of m as:

$$B_n = \frac{2}{L}\int_0^L X_n(x)y_0(x)dx. \tag{6.79}$$

Similarly, to find A_n, we multiply both sides of Eq. (6.76) by $X_m(x)$ and integrate the equation from 0 to L to get:

$$\sum_{n=1}^{\infty} \int_0^L X_m(x)X_n(x)(\omega_n A_n)dx = \int_0^L X_m(x)v_0(x)dx. \tag{6.80}$$

Then using orthogonality,

$$\frac{L}{2}\omega_n A_m = \int_0^L X_m(x)v_0(x)dx, \tag{6.81}$$

and after using subscript n instead of m:

$$A_n = \frac{2}{L\omega_n}\int_0^L X_n(x)v_0(x)dx. \tag{6.82}$$

Then the final solution of a beam will be:

$$y(x,t) = \frac{2}{L}$$
$$\sum_{n=1}^{\infty} X_n(x)\left(\frac{1}{\omega_n}\left(\int_0^L X_n(x)v_0(x)dx\right)\sin\omega_n t + \left(\int_0^L X_n(x)y_0(x)dx\right)\cos\omega_n t\right). \tag{6.83}$$

We should note that it is not practical to use infinite modes. We normally use N modes, with the understanding that by doing so, we introduce an error in the response, due to neglecting all the modes higher than the Nth mode.

6 Traveling Waves and the Dispersion Relation

In this section, we focus on another approach to relate the frequency of oscillations in a beam to its mode shapes. This approach is based on the assumption that the oscillations of a beam (or a string) resemble the motion of a wave that travels from one end of the beam to another end, known as a *traveling wave* response. Sometimes, the motion is a combination of several traveling wave responses. And in some cases, the traveling waves interact in such a way that the response resembles a wave that does not travel, also referred to as a *standing wave*. To discuss more details of this approach, we consider the original form of the equation of motion for a beam with no external tension (Eq. 6.28):

$$\frac{\partial^2 y}{\partial t^2} + c_b^2 \frac{\partial^4 y}{\partial x^4} = 0. \tag{6.84}$$

Then we consider the general form of a traveling wave solution:

$$y(x,t) = Y\sin(\omega t - kx), \tag{6.85}$$

in which ω is the circular frequency of harmonic wave, and $k = 2\pi/\lambda$ is the wave number, where λ is the wavelength. Note that the right-hand side of Eq. (6.85) is a sinusoidal function that depends on both time and space. We can write Eq. (6.85) as:

$$y(x,t) = Y\sin\left(\omega\left(t - \frac{k}{\omega}x\right)\right), \tag{6.86}$$

or

$$y(x,t) = Y\sin\left(\omega\left(t - \frac{x}{c_w}\right)\right), \tag{6.87}$$

where $c_w = \omega/k = 2\pi f/(2\pi/\lambda) = f\lambda$ is called the wave speed.

Equation (6.87) is the equation of a traveling wave that travels from left to right in a horizontal beam. To show this, consider the function $y(x, t)$ at a slightly different time and at a location to the right of its location at time t: $y(x + \Delta x, t + \Delta t)$:

$$\begin{aligned} y(x+\Delta x, t+\Delta t) &= Y \sin\left(\omega\left(t+\Delta t-\frac{x+\Delta x}{c_w}\right)\right) \\ &= Y \sin\left(\omega\left(t-\frac{x}{c_w}+\Delta t-\frac{\Delta x}{c_w}\right)\right). \end{aligned} \tag{6.88}$$

Since the wave speed, c_w, can be written as $c_w = \Delta x/\Delta t$, then $\Delta t - \Delta x/c_w = 0$, and the function becomes:

$$y(x+\Delta x, t+\Delta t) = Y \sin\left(\omega\left(t-\frac{x}{c_w}\right)\right), \tag{6.89}$$

which is the same as $y(x, t)$ in Eq. (6.87). This means that at a Δx location to the right of the original location, and at a time Δt, we will have exactly the same function as $y(x, t)$, and therefore it seems that the function $y(x, t)$ is traveling to the right. Similarly, one can show that

$$y(x, t) = Y \sin(\omega t + kx) \tag{6.90}$$

is a wave that travels from right to left.

If we plug either form of the traveling waves (e.g., Eq. (6.85)) into the equation of motion for a beam (Eq. 6.84), we find:

$$-Y\omega^2 \sin(\omega t - kx) + c_b^2 k^4 Y \sin(\omega t - kx) = 0, \tag{6.91}$$

which is simplified to

$$-\omega^2 + c_b^2 k^4 = 0. \tag{6.92}$$

This equation, also called the *Dispersion Relation*, relates the oscillation frequency, ω, to the "frequency" in space, k. We can refer to k as the "frequency" in space, by comparing the form of the two terms ωt and kx: ω is the frequency in time and k is the frequency in space, which is known as the *wavenumber*. This frequency in space gives us information about the mode that is excited in the beam, since it can be related to the "spatial period", λ, as:

$$\lambda = \frac{2\pi}{k}. \tag{6.93}$$

The "spatial period", λ, is known as the wavelength, which gives the length of one period of the wave. To find the wavelength of a beam, we need its boundary

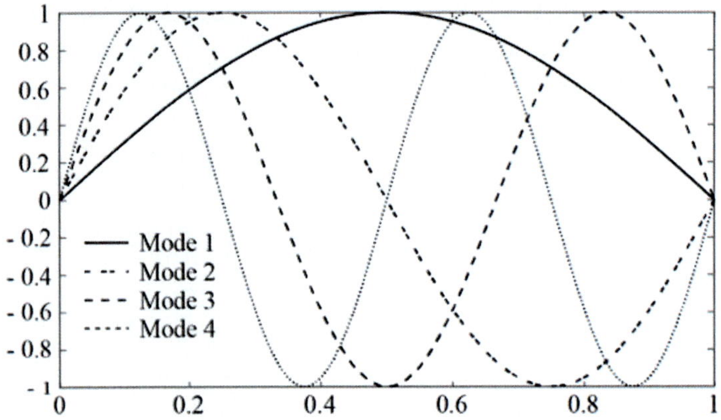

Fig. 6.4 The first four modes of a simply supported beam

conditions, its length, and the mode number. As an example, for the first mode of a simply supported beam, where the mode shape looks like a half sine, the wavelength is 2*L*. For its second mode, the wavelength is *L*, and for the third mode 2*L*/3, and for the *n*-th mode: 2*L*/*n* (Fig. 6.4).

If, for example, mode *n* of a simply supported beam is excited, the wavelength is 2*L*/*n*. We can then use the wavelength and find the wavenumber, since $2L/n = 2\pi/k$ and therefore $k = n\pi/L$. Then, given the dispersion relation for a beam with no external tension:

$$-\omega^2 + c_b^2 k^4 = 0, \tag{6.94}$$

we can calculate the natural frequency as:

$$\omega_n = c_b k_n^2 = c_b \left(\frac{n\pi}{L}\right)^2 = n^2 \pi^2 \frac{c_b}{L^2}, \tag{6.95}$$

and since $c_b = \sqrt{\frac{EI}{\rho A}}$,

$$\omega_n = \frac{n^2 \pi^2}{L^2} \sqrt{\frac{EI}{\rho A}}, \tag{6.96}$$

and since $\omega_n = 2\pi f_n$,

$$f_n = \frac{n^2 \pi}{2L^2} \sqrt{\frac{EI}{\rho A}}. \tag{6.97}$$

Equation (6.97) gives the natural frequencies of different modes of a simply supported beam.

To find the dispersion relation for a beam with external tension, we use the equation of motion of a beam:

$$\rho A \frac{\partial^2 y}{\partial t^2} + EI \frac{\partial^4 y}{\partial x^4} - T \frac{\partial^2 y}{\partial x^2} = f, \tag{6.98}$$

and we assume that there is no external force, $f = 0$, and plug the general form of the traveling wave (Eq. (6.85)) into the equation of motion to get:

$$-\rho A \omega^2 Y \sin(\omega t - kx) + EIk^4 Y \sin(\omega t - kx) + Tk^2 Y \sin(\omega t - kx) = 0, \tag{6.99}$$

or

$$-\rho A \omega^2 + EIk^4 + Tk^2 = 0, \tag{6.100}$$

which relates the frequency in time to the frequency in space:

$$\omega^2 = \frac{EI}{\rho A} k^4 + \frac{T}{\rho A} k^2. \tag{6.101}$$

If we consider this relation for a simply supported beam, then the wave number is

$$k = \frac{2\pi}{\lambda} = \frac{2\pi}{\frac{2L}{n}} = \frac{n\pi}{L}. \tag{6.102}$$

By plugging Eq. (6.102) into the dispersion relation of Eq. (6.101), we get:

$$\omega_n = \sqrt{\left(\frac{n\pi}{L}\right)^4 \frac{EI}{\rho A} + \left(\frac{n\pi}{L}\right)^2 \frac{T}{\rho A}}, \tag{6.103}$$

and since $\omega_n = 2\pi f_n$, the natural frequency of the system in Hz, f_n, can be written as:

$$f_n = \sqrt{\frac{n^2}{4L^2} \left(\frac{n^2\pi^2}{L^2}\right) \frac{EI}{\rho A} + \frac{n^2}{4L^2} \frac{T}{\rho A}}, \tag{6.104}$$

or

$$f_n = \sqrt{\frac{n^4\pi^2}{4L^4} \frac{EI}{\rho A} + \frac{n^2}{4L^2} \frac{T}{\rho A}}. \tag{6.105}$$

In Eq. (6.105), the natural frequencies of a simply supported beam clearly depend on a flexural-rigidity-dependent term (the first term on the right-hand side of Eq. (6.105)) and a tension-dependent term. If one of these two terms is much larger than the other one, then the smaller term can be neglected. In some cases the two terms are comparable and none of them can be neglected, but if the first term is dominating, then the structure acts mainly as a beam, and if the second term is dominating, the structure acts mainly as a string and is called a tension-dominated beam. Therefore, we need to find the ratio of the two terms:

$$R = \frac{\frac{n^4\pi^2}{4L^4}\frac{EI}{\rho A}}{\frac{n^2}{4L^2}\frac{T}{\rho A}} = \frac{n^2\pi^2 EI}{L^2 T}, \tag{6.106}$$

to see if a beam is tension-dominated or not. If $R \gg 1$ then the structure is a flexural-rigidity-dominated beam, and if $R \ll 1$, then the structure is a tension-dominated beam or a string. We should note that this ratio depends on the boundary conditions—the R-value of Eq. (6.106) is based on simply supported boundary conditions. This ratio depends on the system parameters, as expected, and on the mode number, n. As the mode number is increased, the contribution of the flexural-rigidity component is increased, and the structure acts more like a flexural-rigidity-dominated beam.

Another property of the natural frequencies in a simply supported beam is that for a tension-dominated beam, where the first term of Eq. (6.105) is negligible, the natural frequency is written as

$$f_n = \frac{n}{2L}\sqrt{\frac{T}{\rho A}}, \tag{6.107}$$

which means that the natural frequencies of a tension-dominated simply supported beam increase linearly by increasing the mode number. The natural frequency of the 3rd mode is 3 times the natural frequency of the 1st mode and the natural frequency of the 11th mode is 11 times the natural frequency of the 1st mode.

This is different for a flexural-rigidity-dominated beam, for which the natural frequencies are written as

$$f_n = \frac{n^2\pi}{2L^2}\sqrt{\frac{EI}{\rho A}}, \tag{6.108}$$

and therefore, they increase quadratically with increasing mode number. The natural frequency of the 3rd mode is 9 times that of the 1st mode, and the natural frequency of the 11th mode is 121 times that of the 1st mode. This indicates that the natural frequencies are very widely spread at higher modes for a flexural-rigidity-dominated beam compared with a tension-dominated beam. One can then imagine that exciting higher modes are more likely for a tension-dominated beam, since they are much closer to each other compared with a flexural-rigidity-dominated beam (Fig. 6.5).

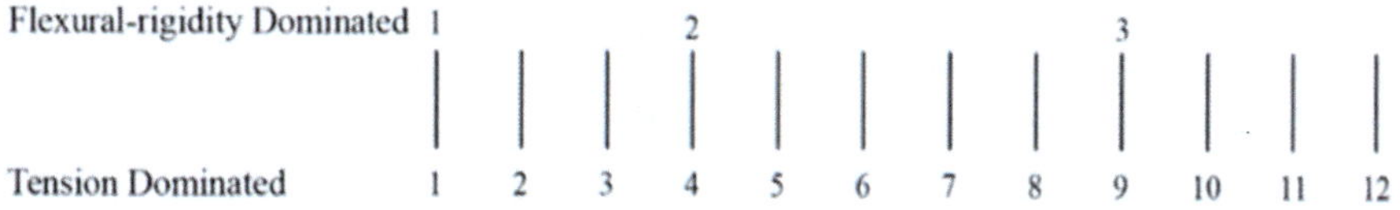

Fig. 6.5 Distribution of natural frequencies for a flexural-rigidity-dominated beam versus a tension-dominated beam

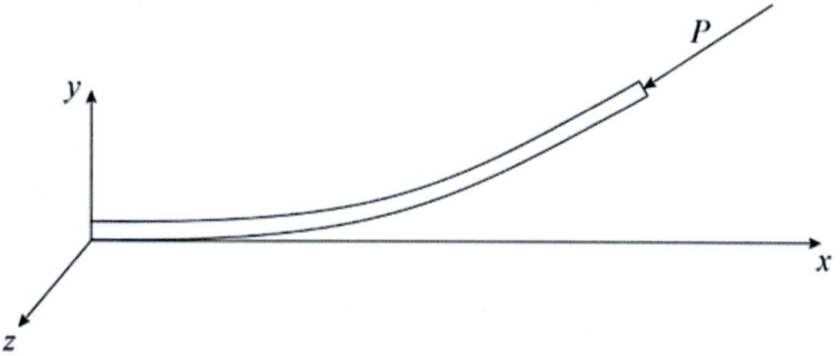

Fig. 6.6 A schematic of a beam with a follower force (the Beck's problem)

7 The Galerkin Method

In this section, we discuss one of the numerical methods that can be used to solve the beam equation: the Galerkin method. We can use the Galerkin method to solve the equations of motion for beams with different boundary conditions, with external forces, and even equations with more terms in them than the terms that exist in the simplest beam equation (Eq. 6.24) that we derived in Sect. 2 of this chapter. The Galerkin method assumes that the response of the system can be written as

$$y(x,t) = \sum_{j=1}^{N} \phi_j(x)\, q_j(t), \tag{6.109}$$

where $\phi_j(x)$ are functions of space only and are called the basis functions, and $q_j(t)$ are functions of time and are called the generalized coordinates. In the case of a beam, the beam's mode shapes, which satisfy its boundary conditions, are used for $\phi_j(x)$, and $q_j(t)$ show the magnitudes of contribution of each mode with time. As we will see later, the Galerkin method converts a partial differential equation (which is the equation that represents the beam) to a set of ordinary differential equations (ODEs), which we will have to solve. The size of this set of ODEs is equal to the number of modes that we use in discretization: N.

To go over the details of the Galerkin method, we consider an example: the Beck's problem. In this problem, a cantilevered beam is under an external pressurization along its length. The external pressurization is applied through a force that acts at the tip of the cantilevered beam, and when the beam bends, the force follows the beam such that its direction always remains the same as the tangent to the tip of the beam. This is also known as the follower force problem—since the force follows the beam.

The equation of motion for this system is very similar to a beam with external tension, but with a positive sign for the y'' term, since instead of a tension force, we have pressurization:

$$m\ddot{y} + EIy'''' + Py'' = 0, \tag{6.110}$$

where $m = \rho A$ is the mass per unit length of the beam. The boundary conditions for a cantilevered beam are:

$$y(0,t) = y'(0,t) = y''(L,t) = y'''(L,t) = 0. \tag{6.111}$$

The goal is to study the beam's response as the external pressurization, P, is increased. We use the Galerkin method to solve this problem. To do so, we assume:

$$y(x,t) = \sum_{j=1}^{N} \phi_j(x)\, q_j(t), \tag{6.112}$$

where N is the number of modes that we use in this discretization. For $\phi_j(x)$, we use the mode shapes of a cantilevered beam.

$$\phi_j(x) = \cosh\beta_j x - \cos\beta_j x - \sigma_j(\sinh\beta_j x - \sin\beta_j x), \tag{6.113}$$

where $\sigma_j = \frac{\cos\beta_j L + \cosh\beta_j L}{\sin\beta_j L + \sinh\beta_j L}$.

By plugging this form into the equation of motion, we get:

$$m\sum_{j=1}^{N} \phi_j(x)\, \ddot{q}_j(t) + EI\sum_{j=1}^{N} \phi_j''''(x)\, q_j(t) + P\sum_{j=1}^{N} \phi_j''(x)\, q_j(t) = 0. \tag{6.114}$$

Note that $(\,)'$ is the derivative with respect to x and $(\,)^{\cdot}$ is the derivative with respect to time. Since only ϕ is a function of x, all derivatives with respect to x appear with a ϕ, and since q is the only function of time, all the dots appear with a q.

The next step in the Galerkin method is to find the contribution of each mode in the final answer. To do so, we multiply Eq. (6.114) by the beam mode shapes, $\phi_r(x)$, to get:

$$m\sum_{j=1}^{N}\phi_r\phi_j\,\ddot{q}_j+EI\sum_{j=1}^{N}\phi_r\phi_j''''\,q_j+P\sum_{j=1}^{N}\phi_r\phi_j''\,q_j=0,\qquad r=1,\ldots,N. \tag{6.115}$$

Note that r varies from 1 to N, because we need to calculate the contributions of all the modes that are used in discretization.

Since the mode shapes cover the entire length of the beam, we integrate Eq. (6.115) over the entire length, i.e., from 0 to L to come up with:

$$\int_0^L\left[m\sum_{j=1}^{N}\phi_r\phi_j\,\ddot{q}_j+EI\sum_{j=1}^{N}\phi_r\phi_j''''\,q_j+P\sum_{j=1}^{N}\phi_r\phi_j''\,q_j\right]dx=0,\qquad r=1,\ldots,N. \tag{6.116}$$

which can be written as:

$$\sum_{j=1}^{N}\left\{m\left[\int_0^L(\phi_r\phi_j)dx\right]\ddot{q}_j+EI\left[\int_0^L\left(\phi_r\phi_j''''\right)dx\right]q_j+P\left[\int_0^L\left(\phi_r\phi_j''\right)dx\right]q_j\right\}=0,\qquad r=1,\ldots,N \tag{6.117}$$

After all the integrals in this equation are calculated, the equation does not depend on x anymore, and it can be written in the following matrix form:

$$m\,\mathbf{A}\,\ddot{\mathbf{q}}+(EI\,\mathbf{B}+P\;\mathbf{C}\,)\mathbf{q}=\mathbf{0}, \tag{6.118}$$

which is in fact in the form of an N-DOF mass-spring system:

$$\mathbf{M}\,\ddot{\mathbf{q}}+\mathbf{K}\,\mathbf{q}=\mathbf{0}. \tag{6.119}$$

In Eqs. (6.118) and (6.119), **A**, **B**, **C**, **M,** and **K** are matrices of the size N × N, and $\ddot{\mathbf{q}}$ and $\mathbf{q}$ are vectors of the size N × 1.

To provide a better understanding of these steps, we consider a simple case where $N = 2$. Then we use the first two mode shapes of a cantilevered beam in our Galerkin discretization: ϕ_1 and ϕ_2. And we will have two unknowns: q_1 and q_2. The solution has the form of

$$\begin{aligned} y(x,t) &= \sum_{j=1}^{2} \phi_j(x)\, q_j(t) \\ &= \phi_1(x) q_1(t) + \phi_2(x) q_2(t). \end{aligned} \tag{6.120}$$

Our set of ODEs (Eq. (6.117)) will have two coupled ODEs. The first ODE will be obtained for $r = 1$ and when j varies from 1 to 2 (from Eq. (6.117)):

$$\begin{aligned} & m\left[\int_0^L (\phi_1\phi_1)dx\right] \ddot{q}_1 + m\left[\int_0^L (\phi_1\phi_2)dx\right] \ddot{q}_2 \\ & + EI\left[\int_0^L (\phi_1\phi_1'''')dx\right] q_1 + EI\left[\int_0^L (\phi_1\phi_2'''')dx\right] q_2 \\ & + P\left[\int_0^L (\phi_1\phi_1'')dx\right] q_1 + P\left[\int_0^L (\phi_1\phi_2'')dx\right] q_2 = 0, \end{aligned} \tag{6.121}$$

and the second ODE is for $r = 2$ and when j varies from 1 to 2:

$$\begin{aligned} & m\left[\int_0^L (\phi_2\phi_1)dx\right] \ddot{q}_1 + m\left[\int_0^L (\phi_2\phi_2)dx\right] \ddot{q}_2 \\ & + EI\left[\int_0^L (\phi_2\phi_1'''')dx\right] q_1 + EI\left[\int_0^L (\phi_2\phi_2'''')dx\right] q_2 \\ & + P\left[\int_0^L (\phi_2\phi_1'')dx\right] q_1 + P\left[\int_0^L (\phi_2\phi_2'')dx\right] q_2 = 0. \end{aligned} \tag{6.122}$$

These two ODEs can be written as:

$$\begin{aligned} & m\, A_{11}\, \ddot{q}_1 + m\, A_{12}\, \ddot{q}_2 \\ & + EI\, B_{11}\, q_1 + EI\, B_{12}\, q_2 \\ & + P\, C_{11}\, q_1 + P\, C_{12}\, q_2 = 0, \end{aligned} \tag{6.123}$$

and

$$\begin{aligned} & m\, A_{21}\, \ddot{q}_1 + m\, A_{22}\, \ddot{q}_2 \\ & + EI\, B_{21}\, q_1 + EI\, B_{22}\, q_2 \\ & + P\, C_{21}\, q_1 + P\, C_{22}\, q_2 = 0, \end{aligned} \tag{6.124}$$

which have the same form as Eq. (6.118).

By using the Galerkin technique, instead of solving the original beam PDE, we need to solve this set of ODEs. The size of this set is equal to the number of modes that we use in the discretization. In this example, we used two modes only. Normally, and in many cases, two modes are not enough. Something very important before we use the Galerkin technique is to make sure that we have used enough modes in our discretization. To do so, we need to do a convergence test: use N modes, solve the problem, get the final answer and then use $N + p$ (where p is a positive integer) modes, solve the problem, get the final answer and compare it with the previous solution until the difference between the two solutions is smaller than an appropriate tolerance.

To solve the set of ODEs of Eq. (6.119), we follow the same steps that we discussed in Chap. 5 to solve a linear 2 degree-of-freedom system. The only difference is that most probably we have more than two equations when we use the Galerkin method, since N is normally larger than 2. To solve this set of equations, we assume

$$\mathbf{q} = \mathbf{A}e^{\lambda t}, \tag{6.125}$$

and plug it into the set of equations to get:

$$\mathbf{M}\lambda^2\mathbf{A}e^{\lambda t} + \mathbf{K}\,\mathbf{A}e^{\lambda t} = \mathbf{0}, \tag{6.126}$$

or

$$\left(\mathbf{M}\,\lambda^2 + \mathbf{K}\right)\mathbf{A}e^{\lambda t} = \mathbf{0}, \tag{6.127}$$

from which we can drop $e^{\lambda t}$ (since it is never zero) to get an eigenvalue problem:

$$\left(\mathbf{M}\,\lambda^2 + \mathbf{K}\right)\mathbf{A} = \mathbf{0}, \tag{6.128}$$

for which to get a nontrivial solution, we need to have:

$$\det\left(\mathbf{M}\,\lambda^2 + \mathbf{K}\right) = 0. \tag{6.129}$$

Note that $\mathbf{M}$ and $\mathbf{K}$ are functions of m, EI, and P. For a known beam, m and EI are known and here we assume that P is increasing from zero to any desirable value. The goal is to investigate the stability of the beam, similar to what we did in Chap. 5 to study the stability of a 2 DOF airfoil.

For the sake of simplicity, we use only two modes and calculate the eigenvalues for increasing P. The plots of Fig. 6.7 show the real and imaginary parts of the eigenvalues. We remember from Chap. 5 that the imaginary part of the eigenvalue is proportional to the frequency of the response and the real part of the eigenvalue is

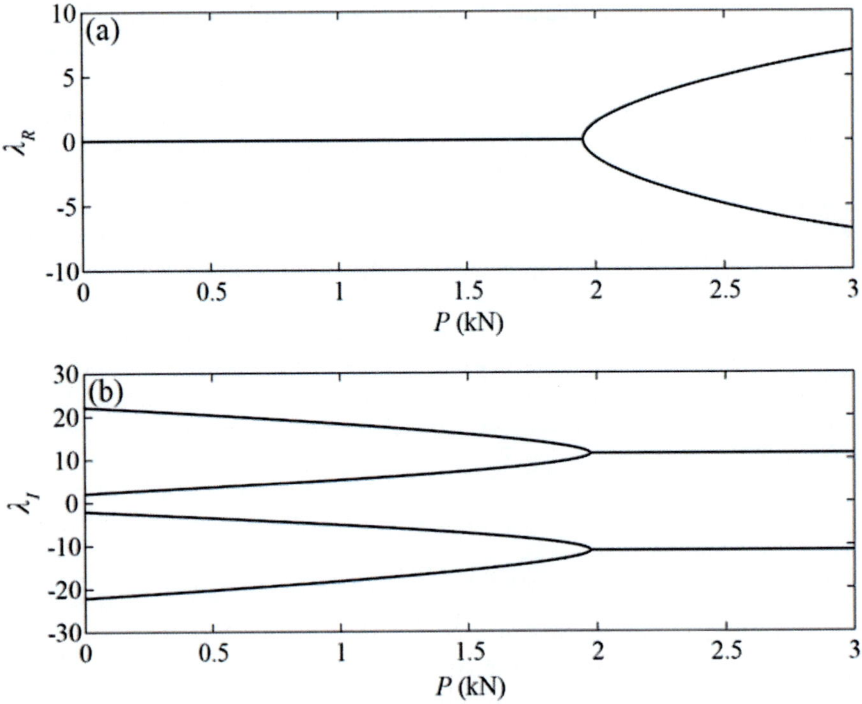

Fig. 6.7 (**a**) Real and (**b**) imaginary parts of the eigenvalues for a follower force problem. The results are for a steel beam of a circular cross section with a diameter of 1 cm

proportional to the negative damping of the system. The plots of Fig. 6.7 show that as P increases, the two beam modes coalesce at a critical value of P, and at the same critical value, a negative damping is observed, resulting in a coupled-mode flutter.

Chapter 7
A Flexible Pipe Conveying Fluid

1 Introduction

In this chapter and the following chapter, we will discuss a class of FSI problems in which a relatively long flexible structure is placed in flow such that the direction of flow is in the direction of the long axis of the structure. Several examples of these problems exist. When a flag flaps in wind, the wind is mainly parallel to its long axis. In our bodies, blood moves in parallel to the walls of the artery. We call these problems axial flow problems, because the flow is in the direction of the long axis of the structure. Typical structures that undergo axial FSI are flexible beams, strings, thin plates, thin shells, and membranes, placed in flow. In axial flow problems, the flow could be flowing inside the structure—such as blood inside the artery—or outside the structure—such as the flag problem. These two types of problems are referred to as internal and external axial flow problems, respectively. The most common example of an internal axial flow problem is a flexible pipe conveying fluid, which is what we will discuss in this chapter, and its counterpart for external axial flow problems, is a flexible cylinder placed in axial flow, which we will study in the next chapter. The pipe problem is probably the most extensively studied axial flow problem. There are several reasons for popularity of the pipe problem: the equation that represents the system is relatively easy, it is relatively easy to conduct experimental studies on a pipe conveying fluid, and although the system is relatively simple, very rich dynamics can be observed in it. The pipe problem has been studied extensively to understand concepts in structural stability, nonlinear dynamics, and chaos.

In this chapter, we will discuss both linear and nonlinear models for a pipe conveying fluid, together with methods of solving these equations, and we will get an overview of different types of instability that a pipe can undergo, and the types of response that are observed after these instabilities occur. We will see how depending on the boundary conditions, the system properties, and the flow velocity, pipes

Y. Modarres-Sadeghi, *Introduction to Fluid-Structure Interactions*,
https://doi.org/10.1007/978-3-030-85884-1_7

conveying fluid can undergo both static and dynamic instabilities, and can exhibit limit cycle oscillations, period-doubling, quasiperiodicity, and chaos.

2 Derivation of a Linear Equation for a Pipe Conveying Fluid

A pipe conveying fluid is a simple system, which is used as a model problem to study several generic characteristics of dynamical systems. Different types of dynamical responses can be observed by simple adjustments of the system, i.e., by attaching a mass or a spring to it, or by changing its boundary conditions.

To derive the equation of motion for a pipe conveying fluid, we consider an arbitrary element of the pipe with a width of δx placed at a distance x from one end of the pipe (Fig. 7.1). This element consists of a beam (the pipe) and the fluid that is inside the element. We consider the free-body diagrams for each of these two parts of the element as shown in Fig. 7.2. We assume that the pipe oscillates in the y-direction only (the direction perpendicular to the long axis of the pipe), and its displacements in the x-direction (the direction of the long axis of the pipe at its original equilibrium position) are negligible compared with those in the y-direction. To quantify this assumption, we say oscillations in the x-direction are an order of magnitude smaller than those in the y-direction. Then we derive a linear equation of motion by

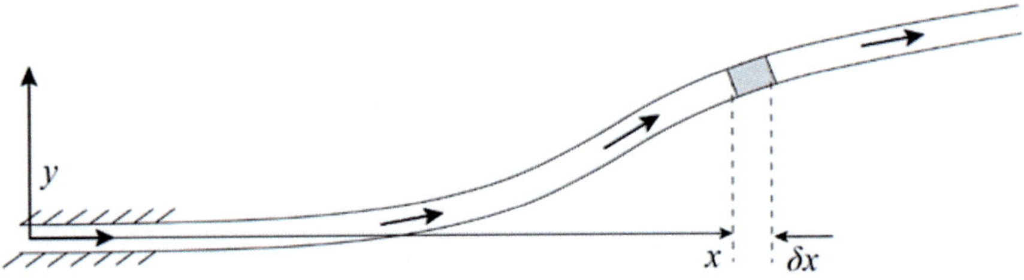

Fig. 7.1 A schematic of a pipe conveying fluid

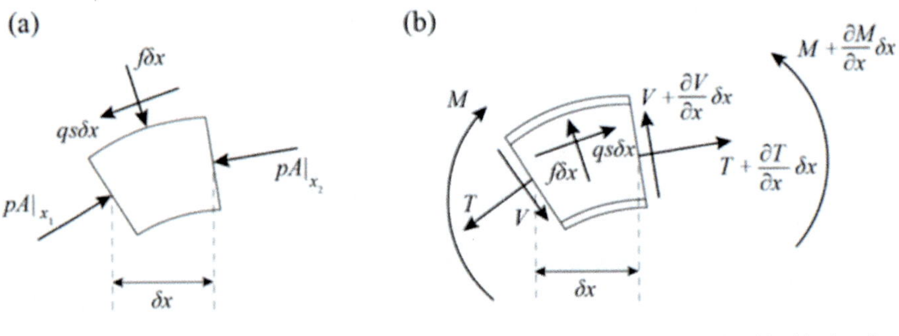

Fig. 7.2 Free-body diagram of an element along the length of the pipe: (**a**) the fluid inside the pipe and (**b**) the structure

considering the free-body diagrams of Fig. 7.2. The forces that act on the pipe are the same as the forces that act on a beam in transverse oscillations, as we discussed in Chap. 6, plus two external forces that act on the pipe due to the flow: $f\delta x$ and $qs\delta x$, where f is the flow force per unit length that acts perpendicular to the pipe, similar to the general distributed external force that acted on a beam, as we discussed in Sect. 6, and q is the force per unit area that acts on the pipe in the flow direction.

First, we consider the free-body diagram of the fluid inside the pipe. Then we will have:

$$\begin{aligned} &\sum f_x = ma_x \\ &\rightarrow +pA|_{x_1} - pA|_{x_2} - qs\delta x + f\delta x\frac{\partial y}{\partial x} = 0 \end{aligned} \tag{7.1}$$

and

$$\begin{aligned} &\sum f_y = ma_y \\ &\rightarrow pA\frac{\partial y}{\partial x}\bigg|_{x_1} - pA\frac{\partial y}{\partial x}\bigg|_{x_2} - qs\delta x\frac{\partial y}{\partial x} - f\delta x = M\delta x a_{fy} \end{aligned} \tag{7.2}$$

where, in both of these equations, M is the mass of the fluid inside the pipe per unit length. Then to relate $pA|_{x_2}$ to $pA|_{x_1}$, we use the Taylor expansion to get:

$$pA|_{x_2} = pA|_{x_1} + \frac{\partial}{\partial x}(pA)\delta x, \tag{7.3}$$

and by assuming that the cross-section remains constant:

$$pA|_{x_2} = pA|_{x_1} + A\frac{\partial p}{\partial x}\delta x, \tag{7.4}$$

and therefore Eq. (7.1) becomes:

$$pA|_{x_1} - pA|_{x_1} - A\frac{\partial p}{\partial x}\delta x - qs\delta x + f\delta x\frac{\partial y}{\partial x} = 0 \tag{7.5}$$

or

$$-A\frac{\partial P}{\partial x}\delta x - qs\delta x + f\delta x\frac{\partial y}{\partial x} = 0, \tag{7.6}$$

and by dropping δx from all terms:

$$-A\frac{\partial P}{\partial x} - qs + f\frac{\partial y}{\partial x} = 0. \tag{7.7}$$

Similarly, to relate $pA\frac{\partial y}{\partial x}\Big|_{x_2}$ to $pA\frac{\partial y}{\partial x}\Big|_{x_1}$ in Eq. (7.2), we have

$$pA\frac{\partial y}{\partial x}\Big|_{x_2} = pA\frac{\partial y}{\partial x}\Big|_{x_1} + \frac{\partial}{\partial x}\left(pA\frac{\partial y}{\partial x}\right)\delta x, \tag{7.8}$$

which makes Eq. (7.2):

$$pA\frac{\partial y}{\partial x}\Big|_{x_1} - pA\frac{\partial y}{\partial x}\Big|_{x_1} - \frac{\partial}{\partial x}\left(pA\frac{\partial y}{\partial x}\right)\delta x - qs\delta x\frac{\partial y}{\partial x} - f\delta x = M\delta x a_{fy}, \tag{7.9}$$

or

$$-\frac{\partial}{\partial x}\left(pA\frac{\partial y}{\partial x}\right)\delta x - qs\delta x\frac{\partial y}{\partial x} - f\delta x = M\delta x a_{fy}, \tag{7.10}$$

and after dropping δx from all terms:

$$-\frac{\partial}{\partial x}\left(pA\frac{\partial y}{\partial x}\right) - qs\frac{\partial y}{\partial x} - f = Ma_{fy}. \tag{7.11}$$

The acceleration that we observe on the right-hand side of Eq. (7.11) is the acceleration of the fluid element, and in order to find it, we need to calculate the second derivative of the fluid's position, $y(x(t), t)$ with respect to time. We start by finding the velocity:

$$v = \frac{\partial}{\partial t}y(x(t), t) = \frac{\partial y}{\partial x}\frac{\partial x}{\partial t} + \frac{\partial y}{\partial t}, \tag{7.12}$$

where, we have used the chain rule, because x is also a function of t. What we have found on the right-hand side of Eq. (7.12) is what is called the *material derivate* of y, also written as:

$$\frac{Dy}{Dt} = \frac{\partial y}{\partial t} + \frac{\partial y}{\partial x}\frac{\partial x}{\partial t}, \tag{7.13}$$

which is used when we find the derivate with respect to time of a quantity that itself is a function of variables that are functions of time! In Eq. (7.13), $\frac{\partial x}{\partial t}$ is the fluid velocity in the x-direction, which is equal to the flow velocity inside the pipe. Therefore, Eq. (7.12) can be written as:

$$v = \frac{Dy}{Dt} = \frac{\partial y}{\partial t} + U\frac{\partial y}{\partial x}, \tag{7.14}$$

To find the acceleration we find the material derivative of the velocity:

$$\begin{aligned} a &= \frac{D}{Dt}v = \left(\frac{\partial}{\partial t} + U\frac{\partial}{\partial x}\right)\left(\frac{\partial y}{\partial t} + U\frac{\partial y}{\partial x}\right) \\ &= \left(\frac{\partial^2 y}{\partial t^2} + U\frac{\partial^2 y}{\partial x \partial t}\right) + U\left(\frac{\partial^2 y}{\partial x \partial t} + U\frac{\partial^2 y}{\partial x^2}\right) \\ &= \left(\frac{\partial^2 y}{\partial t^2} + 2U\frac{\partial^2 y}{\partial x \partial t} + U^2\frac{\partial^2 y}{\partial x^2}\right). \end{aligned} \tag{7.15}$$

and therefore:

$$a_{fy} = \left(\frac{\partial^2 y}{\partial t^2} + 2U\frac{\partial^2 y}{\partial x \partial t} + U^2\frac{\partial^2 y}{\partial x^2}\right). \tag{7.16}$$

Now that we have the acceleration, we can update Eq. (7.11) to get

$$-\frac{\partial}{\partial x}\left(pA\frac{\partial y}{\partial x}\right) - qs\frac{\partial y}{\partial x} - f = M\left(\frac{\partial^2 y}{\partial t^2} + 2U\frac{\partial^2 y}{\partial x \partial t} + U^2\frac{\partial^2 y}{\partial x^2}\right). \tag{7.17}$$

Then Eqs. (7.7) and (7.17) are the two equations that we obtain by considering the free-body diagram of the fluid only.

Now, we consider the free-body diagram of the structure. This derivation is very similar to the derivation of the equation of motion for a general linear beam as we discussed in Chap. 6. First, in the x-direction we get:

$$\sum F_x = 0 \Rightarrow -f\delta x\frac{\partial y}{\partial x} + qS\delta x - T + T + \frac{\partial T}{\partial x}\delta x + V\frac{\partial y}{\partial x} - \left(V + \frac{\partial V}{\partial x}\delta x\right)\frac{\partial y}{\partial x} = 0, \tag{7.18}$$

which can be simplified to

$$-f\delta x\frac{\partial y}{\partial x} + qS\delta x + \frac{\partial T}{\partial x}\delta x - \frac{\partial V}{\partial x}\frac{\partial y}{\partial x}\delta x = 0, \tag{7.19}$$

and since V is a function of y ($V = -EI\,\partial^3 y/\partial x^3$), the last term is of the second order and can be neglected, which leaves us with the following equation in the x-direction:

$$-f\frac{\partial y}{\partial x} + qS + \frac{\partial T}{\partial x} = 0. \tag{7.20}$$

Then in the y-direction:

$$\sum F_y = ma_y \Rightarrow -V + \left(V + \frac{\partial V}{\partial x}\delta x\right) + f\delta x + qS\delta x\frac{\partial y}{\partial x} - T\frac{\partial y}{\partial x}\bigg|_{x_1} + T\frac{\partial y}{\partial x}\bigg|_{x_2}$$
$$= m\delta x\frac{\partial^2 y}{\partial t^2}, \tag{7.21}$$

and[1] we relate $T\frac{\partial y}{\partial x}\big|_{x_2}$ to $T\frac{\partial y}{\partial x}\big|_{x_1}$ as

$$T\frac{\partial y}{\partial x}\bigg|_{x_2} = T\frac{\partial y}{\partial x}\bigg|_{x_1} + \frac{\partial}{\partial x}\left(T\frac{\partial y}{\partial x}\right)\delta x, \tag{7.22}$$

and plug it back into Eq. (7.21) to get:

$$\frac{\partial V}{\partial x}\delta x + f\delta x + qS\delta x\frac{\partial y}{\partial x} + \frac{\partial}{\partial x}\left(T\frac{\partial y}{\partial x}\right)\delta x = m\delta x\frac{\partial^2 y}{\partial t^2}, \tag{7.23}$$

or

$$\frac{\partial V}{\partial x} + f + qS\frac{\partial y}{\partial x} + \frac{\partial}{\partial x}\left(T\frac{\partial y}{\partial x}\right) = m\frac{\partial^2 y}{\partial t^2}, \tag{7.24}$$

and since $V = -EI\frac{\partial^3 y}{\partial x^3}$ and $m = \rho A$,

$$-\frac{\partial^2\left(EI\frac{\partial^2 y}{\partial x^2}\right)}{\partial x^2} + f + qS\frac{\partial y}{\partial x} + \frac{\partial}{\partial x}\left(T\frac{\partial y}{\partial x}\right) = \rho A\frac{\partial^2 y}{\partial t^2}, \tag{7.25}$$

and for a constant *EI*:

$$-EI\frac{\partial^4 y}{\partial x^4} + f + qS\frac{\partial y}{\partial x} + \frac{\partial}{\partial x}\left(T\frac{\partial y}{\partial x}\right) = \rho A\frac{\partial^2 y}{\partial t^2}. \tag{7.26}$$

Overall, we obtain 4 equations from the two free-body diagrams:

$$-A\frac{\partial p}{\partial x} - qs + f\frac{\partial y}{\partial x} = 0, \tag{7.27}$$

[1]Note that although at x_2 the tension is $T + \frac{\partial T}{\partial x}\delta x$, when we calculate the slope, $\left(T + \frac{\partial T}{\partial y}\delta x\right)\partial y\frac{\partial y}{\partial x} = T\frac{\partial y}{\partial x}$, since the second term is of the second order.

$$-A\frac{\partial}{\partial x}\left(p\frac{\partial y}{\partial x}\right) - qs\frac{\partial y}{\partial x} - f = M\left(\frac{\partial^2 y}{\partial t^2} + 2U\frac{\partial^2 y}{\partial x \partial t} + U^2\frac{\partial^2 y}{\partial x^2}\right) \quad (7.28)$$

$$-f\frac{\partial y}{\partial x} + qS + \frac{\partial T}{\partial x} = 0, \quad (7.29)$$

$$-EI\frac{\partial^4 y}{\partial x^4} + f + qS\frac{\partial y}{\partial x} + \frac{\partial}{\partial x}\left(T\frac{\partial y}{\partial x}\right) = m\frac{\partial^2 y}{\partial t^2}, \quad (7.30)$$

and we are looking for one equation in terms of y, since y is the only unknown. To find our equation of motion, we first add Eqs. (7.28) and (7.30) to get:

$$\begin{aligned}-EI\frac{\partial^4 y}{\partial x^4} + f + qS\frac{\partial y}{\partial x} + \frac{\partial}{\partial x}\left(T\frac{\partial y}{\partial x}\right) - A\frac{\partial}{\partial x}\left(p\frac{\partial y}{\partial x}\right) - qs\frac{\partial y}{\partial x} - f \\ = m\frac{\partial^2 y}{\partial t^2} + M\left(\frac{\partial^2 y}{\partial t^2} + 2U\frac{\partial^2 y}{\partial x \partial t} + U^2\frac{\partial^2 y}{\partial x^2}\right),\end{aligned} \quad (7.31)$$

which can be simplified to:

$$\begin{aligned}-EI\frac{\partial^4 y}{\partial x^4} + \frac{\partial}{\partial x}\left(T\frac{\partial y}{\partial x}\right) - A\frac{\partial}{\partial x}\left(p\frac{\partial y}{\partial x}\right) \\ = m\frac{\partial^2 y}{\partial t^2} + M\left(\frac{\partial^2 y}{\partial t^2} + 2U\frac{\partial^2 y}{\partial x \partial t} + U^2\frac{\partial^2 y}{\partial x^2}\right),\end{aligned} \quad (7.32)$$

and re-written as:

$$EI\frac{\partial^4 y}{\partial x^4} - \frac{\partial}{\partial x}\left((T - pA)\frac{\partial y}{\partial x}\right) + m\frac{\partial^2 y}{\partial t^2} + M\left(\frac{\partial^2 y}{\partial t^2} + 2U\frac{\partial^2 y}{\partial x \partial t} + U^2\frac{\partial^2 y}{\partial x^2}\right) = 0. \quad (7.33)$$

This equation seems to be very close to the equation that we are looking for. If we now add Eqs. (7.27) and (7.29), we get:

$$-A\frac{\partial p}{\partial x} - qs + f\frac{\partial y}{\partial x} - f\frac{\partial y}{\partial x} + qS + \frac{\partial T}{\partial x} = 0 \quad (7.34)$$

or

$$-A\frac{\partial p}{\partial x}+\frac{\partial T}{\partial x}=0, \tag{7.35}$$

which can be re-written as

$$\frac{\partial}{\partial x}(T-pA)=0. \tag{7.36}$$

Equation (7.36) implies that *T–pA* does not change along the length of the pipe. If we plug this equation into Eq. (7.33), then we will get

$$EI\frac{\partial^4 y}{\partial x^4}-(T-pA)\frac{\partial^2 y}{\partial x^2}+m\frac{\partial^2 y}{\partial t^2}+M\left(\frac{\partial^2 y}{\partial t^2}+2U\frac{\partial^2 y}{\partial x\partial t}+U^2\frac{\partial^2 y}{\partial x^2}\right)=0, \tag{7.37}$$

and after collecting the terms:

$$EI\frac{\partial^4 y}{\partial x^4}+\left(MU^2-(T-pA)\right)\frac{\partial^2 y}{\partial x^2}+2MU\frac{\partial^2 y}{\partial x\partial t}+(m+M)\frac{\partial^2 y}{\partial t^2}=0. \tag{7.38}$$

Eq. (7.38) is the equation of motion for a pipe conveying fluid under external tension and pressurization. We already know that *T–pA* does not change along the length of the pipe. We will make one more assumption and that is $T–pA = 0$, which means that there is no external tension or pressurization acting on the pipe. By making this assumption, we come up with the simplest possible form of an equation of motion for a pipe conveying fluid:

$$EI\frac{\partial^4 y}{\partial x^4}+MU^2\frac{\partial^2 y}{\partial x^2}+2MU\frac{\partial^2 y}{\partial x\partial t}+(m+M)\frac{\partial^2 y}{\partial t^2}=0. \tag{7.39}$$

In this equation, *EI* is the flexural rigidity, *M* is mass of the fluid inside the pipe per unit length, *m* is mass per unit length of the pipe, and *U* is the flow velocity.

The four terms in Eq. (7.39) are, respectively, the flexural restoring force, centripetal force, Coriolis force, and inertial force. There are three terms that are due to the flow: $MU^2\partial^2y/\partial x^2$, $2MU\partial^2y/\partial x\partial t$, and $M\partial^2y/\partial t^2$. The last one is the mass that is added to the system due to the presence of the fluid inside the pipe, and acts as an added mass term in the inertia force. The term $MU^2\partial^2y/\partial x^2$ is a centripetal force in which $\partial^2y/\partial x^2 = 1/R$, where *R* is the radius of curvature of the pipe, and therefore $MU^2\partial^2y/\partial x^2 = MU^2/R$, which is the general form of a centripetal force. The term $2MU\partial^2y/\partial x\partial t$ is a Coriolis force in which $\partial y/\partial x \equiv \theta$ and therefore $2MU\partial^2y/\partial x\partial t = 2MU\partial\theta/\partial t = 2MU\dot{\theta}$, which is the general form of a Coriolis force in which a mass *M* moves with a translational velocity *U* along the length of an object that rotates with an angular velocity $\dot{\theta}$.

This is the simplest form of the equation of motion for a pipe conveying fluid, and clearly, it is a linear equation. Note that we have assumed that gravity does not act, and that there is no structural damping, external pressure, or tension. Adding these effects will add more terms to this equation; nevertheless, Eq. (7.39) is the core of the equation of motion for a pipe conveying fluid. Using this simple equation, we can understand several features that are observed in the response of a pipe conveying fluid.

3 Dimensionless Equation and Dimensionless Parameters

Before solving the equation of motion and studying the physics of a pipe conveying fluid, we make the equation dimensionless. Equation (7.39) can be written as:

$$\frac{\partial^4 y}{\partial x^4} + \frac{MU^2}{EI}\frac{\partial^2 y}{\partial x^2} + 2\frac{MU}{EI}\frac{\partial^2 y}{\partial x \partial t} + \frac{(m+M)}{EI}\frac{\partial^2 y}{\partial t^2} = 0. \tag{7.40}$$

To make the equation dimensionless, we need to decide on a dimensionless form for the unknown, y, and the dimensionless forms for the independent variables, x and t. To make the lengths, y and x, dimensionless, we divide them by the length of the pipe and define our dimensionless lengths as:

$$\eta = \frac{y}{L}, \tag{7.41}$$

and

$$\xi = \frac{x}{L}. \tag{7.42}$$

In order to find a dimensionless time, we need to divide time by a constant with a unit of time:

$$\tau = \frac{t}{C}. \tag{7.43}$$

For now, we keep this constant as C, but we keep in mind that we will have to relate it to the system parameters in the process of nondimensionalization. After making these assumptions, we can write the equation of motion as:

$$\frac{\partial^4 (\eta L)}{\partial (\xi L)^4} + \frac{MU^2}{EI}\frac{\partial^2 (\eta L)}{\partial (\xi L)^2} + 2\frac{MU}{EI}\frac{\partial^2 (\eta L)}{\partial (\xi L)\partial (\tau C)} + \frac{(m+M)}{EI}\frac{\partial^2 (\eta L)}{\partial (\tau C)^2} = 0, \tag{7.44}$$

or

$$\frac{L}{L^4}\frac{\partial^4\eta}{\partial\xi^4}+\frac{MU^2L}{EIL^2}\frac{\partial^2\eta}{\partial\xi^2}+2\frac{MUL}{EILC}\frac{\partial^2\eta}{\partial\xi\partial\tau}+\frac{(m+M)L}{EIC^2}\frac{\partial^2\eta}{\partial\tau^2}=0, \tag{7.45}$$

and after some simplifications:

$$\frac{1}{L^3}\frac{\partial^4\eta}{\partial\xi^4}+\frac{MU^2}{EIL}\frac{\partial^2\eta}{\partial\xi^2}+2\frac{MU}{EIC}\frac{\partial^2\eta}{\partial\xi\partial\tau}+\frac{(m+M)L}{EIC^2}\frac{\partial^2\eta}{\partial\tau^2}=0. \tag{7.46}$$

Then we multiply all terms by L^3 to get:

$$\frac{\partial^4\eta}{\partial\xi^4}+\frac{MU^2L^2}{EI}\frac{\partial^2\eta}{\partial\xi^2}+2\frac{MUL^3}{EIC}\frac{\partial^2\eta}{\partial\xi\partial\tau}+\frac{(m+M)L^4}{EIC^2}\frac{\partial^2\eta}{\partial\tau^2}=0. \tag{7.47}$$

Note that the first term in this equation is dimensionless: it is the fourth derivative of a dimensionless parameter with respect to another dimensionless parameter. This means that all the other terms must be dimensionless as well, because they have been added to a dimensionless term. Therefore, the last term is also dimensionless, i.e.,

$$\left[\frac{(m+M)L^4}{EIC^2}\frac{\partial^2\eta}{\partial\tau^2}\right]=1, \tag{7.48}$$

from which one can write:

$$\left[\frac{\partial^2\eta}{\partial\tau^2}\right]=\left[\frac{EIC^2}{(m+M)L^4}\right]. \tag{7.49}$$

Then, since the left-hand side of this equation is dimensionless, its right-hand side must be dimensionless too:

$$\left[\frac{EIC^2}{(m+M)L^4}\right]=1, \tag{7.50}$$

from where we can find the constant C, which we introduced initially in our dimensionless time, as:

$$C=\left(\frac{m+M}{EI}\right)^{1/2}L^2, \tag{7.51}$$

and as a result, the dimensionless time will be:

$$\tau = \left(\frac{EI}{m+M}\right)^{½} \frac{1}{L^2} t. \tag{7.52}$$

Then we plug Eq. (7.51) into Eq. (7.47) to get

$$\frac{\partial^4 \eta}{\partial \xi^4} + \frac{MU^2L^2}{EI} \frac{\partial^2 \eta}{\partial \xi^2} + 2 \frac{MUL^3}{EI\left(\frac{m+M}{EI}\right)^{½} L^2} \frac{\partial^2 \eta}{\partial \xi \partial \tau} + \frac{(m+M)L^4}{EI\left(\frac{m+M}{EI}\right) L^4} \frac{\partial^2 \eta}{\partial \tau^2} = 0, \tag{7.53}$$

and simplify it to:

$$\frac{\partial^4 \eta}{\partial \xi^4} + \frac{MU^2L^2}{EI} \frac{\partial^2 \eta}{\partial \xi^2} + 2 \frac{MUL}{(EI(m+M))^{½}} \frac{\partial^2 \eta}{\partial \xi \partial \tau} + \frac{\partial^2 \eta}{\partial \tau^2} = 0. \tag{7.54}$$

In this version of the equation of motion, coefficients of the first and the last terms are 1 and the coefficients of the second and the third terms are dimensionless. If we write these two coefficients in terms of two dimensionless parameters, then the system's behavior will depend on those two parameters only. The coefficient of the second term has the square of the flow velocity in its numerator. Inspired by the presence of this flow velocity in that term, we define a dimensionless flow velocity, u, as:

$$u = \left(\frac{M}{EI}\right)^{½} UL, \tag{7.55}$$

and as a result, the coefficient of the second term becomes u^2. Then we write the coefficient of the third term as:

$$\begin{aligned}\frac{MUL}{(EI(m+M))^{½}} &= \frac{M^{½}M^{½}UL}{(EI)^{½}(m+M)^{½}} = \left(\frac{M}{EI}\right)^{½} UL \frac{M^{½}}{(m+M)^{½}} \\ &= u \frac{M^{½}}{(m+M)^{½}},\end{aligned} \tag{7.56}$$

in which a dimensionless flow velocity is observed multiplied by a second dimensionless coefficient, $M^{1/2}/(m+M)^{1/2}$. This coefficient relates the mass of the fluid inside the pipe, M, to the total mass, $m+M$. Inspired by this coefficient, we define a new parameter, β, which is a mass ratio parameter as:

$$\beta = \frac{M}{m+M}. \tag{7.57}$$

Then using u and β, we can write Eq. (7.54) as:

$$\frac{\partial^4 \eta}{\partial \xi^4} + u^2 \frac{\partial^2 \eta}{\partial \xi^2} + 2u\beta^{1/2} \frac{\partial^2 \eta}{\partial \xi \partial \tau} + \frac{\partial^2 \eta}{\partial \tau^2} = 0, \tag{7.58}$$

which is the dimensionless form of the simplest equation of motion for a pipe conveying fluid. From this equation, it becomes clear that a pipe's behavior depends on two parameters: u and β. We should note that β is known when a pipe is selected. By selecting a pipe, we know its mass per unit length, m, and its inner diameter, which will give us the mass of the fluid inside the pipe, M, with the assumption that we know what type of fluid the pipe will convey. Then the entire behavior of the pipe will depend on u—the dimensionless flow velocity. We should remember that since we are solving a PDE, when we pick a pipe, we have to pick its boundary conditions as well. As we will see in the following sections, the boundary conditions influence the pipe's behavior significantly.

4 The Method of Solution for the Linear Pipe Equation

To understand the behavior of a pipe conveying fluid as the flow velocity is increased, we first consider the dimensionless version of the equation of motion, in its simplest form:

$$\frac{\partial^4 \eta}{\partial \xi^4} + u^2 \frac{\partial^2 \eta}{\partial \xi^2} + 2u\beta^{1/2} \frac{\partial^2 \eta}{\partial \xi \partial \tau} + \frac{\partial^2 \eta}{\partial \tau^2} = 0. \tag{7.59}$$

To solve this equation, we use the Galerkin technique, as we discussed in Chap. 6, to discretize the partial differential equation. To do so, we assume the following solution:

$$\eta(\xi, \tau) = \sum_{j=1}^{N} \phi_j(\xi)\, q_j(\tau), \tag{7.60}$$

in which $\phi_j(\xi)$ are mode shapes of a beam with the boundary conditions of the pipe, $q_j(\tau)$ are the generalized coordinates, and N is the number of modes that are used in discretization. Then we plug the solution (Eq. (7.60)) into the linear equation of motion (Eq. (7.59)) to get:

$$\sum_{j=1}^{N} \phi_j''''(\xi)\, q_j(\tau) + u^2 \sum_{j=1}^{N} \phi_j''(\xi)\, q_j(\tau) + 2u\beta^{1/2} \sum_{j=1}^{N} \phi_j'(\xi)\, \dot{q}_j(\tau) + \sum_{j=1}^{N} \phi_j(\xi)\, \ddot{q}_j(\tau) = 0. \tag{7.61}$$

Then in order to calculate the contribution of each mode to the total solution, we multiply this equation by $\phi_i(\xi)$ and integrate it over the entire length of the beam (which in its dimensionless form is $\xi = 1$), and we get:

$$\sum_{j=1}^{N}\left(\int_0^1 \phi_i(\xi)\phi_j''''(\xi)\mathrm{d}\xi\right) q_j(\tau) + u^2\sum_{j=1}^{N}\left(\int_0^1 \phi_i(\xi)\phi_j''(\xi)\mathrm{d}\xi\right) q_j(\tau)$$
$$+2u\beta^{1/2}\sum_{j=1}^{N}\left(\int_0^1 \phi_i(\xi)\phi_j'(\xi)\mathrm{d}\xi\right) \dot{q}_j(\tau) + \sum_{j=1}^{N}\left(\int_0^1 \phi_i(\xi)\phi_j(\xi)\mathrm{d}\xi\right) \ddot{q}_j(\tau) = 0, \tag{7.62}$$

which can be re-written as:

$$\sum_{j=1}^{N} A_{1,ij}\, q_j(\tau) + u^2\sum_{j=1}^{N} A_{2,ij}\, q_j(\tau) + 2u\beta^{1/2}\sum_{j=1}^{N} A_{3,ij}\, \dot{q}_j(\tau) + \sum_{j=1}^{N} A_{4,ij}\, \ddot{q}_j(\tau) = 0, \tag{7.63}$$

or

$$\left(\sum_{j=1}^{N} A_{1,ij} + u^2\sum_{j=1}^{N} A_{2,ij}\right) q_j(\tau) + 2u\beta^{1/2}\sum_{j=1}^{N} A_{3,ij}\, \dot{q}_j(\tau) + \sum_{j=1}^{N} A_{4,ij}\, \ddot{q}_j(\tau) = 0 \tag{7.64}$$

where

$$\begin{aligned} A_{1,ij} &= \textstyle\int_0^1 \phi_i(\xi)\phi_j''''(\xi)\mathrm{d}\xi, \\ A_{2,ij} &= \textstyle\int_0^1 \phi_i(\xi)\phi_j''(\xi)\mathrm{d}\xi, \\ A_{3,ij} &= \textstyle\int_0^1 \phi_i(\xi)\phi_j'(\xi)\mathrm{d}\xi, \\ A_{4,ij} &= \textstyle\int_0^1 \phi_i(\xi)\phi_j(\xi)\mathrm{d}\xi. \end{aligned} \tag{7.65}$$

Then Eq. (7.64) can be written in the form of a mass-spring-damper system with N degrees of freedom:

$$K_{ij}\, q_j(\tau) + C_{ij}\, \dot{q}_j(\tau) + M_{ij}\, \ddot{q}_j(\tau) = 0, \tag{7.66}$$

where

$$K_{ij} = \sum_{j=1}^{N} A_{1,ij} + u^2\sum_{j=1}^{N} A_{2,ij}, \tag{7.67}$$

$$C_{ij} = 2u\beta^{1/2}\sum_{j=1}^{N} A_{3,ij}, \tag{7.68}$$

and

$$M_{ij} = \sum_{j=1}^{N} A_{4,ij}, \tag{7.69}$$

are the linear stiffness, damping, and mass matrices, respectively.

In order to solve Eq. (7.66), we re-write it in its first-order form, by assuming

$$\dot{q}_j(\tau) = p_j(\tau). \tag{7.70}$$

This assumption turns Eq. (7.66) into

$$K_{ij}\, q_j(\tau) + C_{ij}\, p_j(\tau) + M_{ij}\, \dot{p}_j(\tau) = 0, \tag{7.71}$$

which can be written in terms of $\dot{p}_j(\tau)$:

$$\dot{p}_j(\tau) = -M_{ij}^{-1} K_{ij}\, q_j(\tau) - M_{ij}^{-1} C_{ij}\, p_j(\tau), \tag{7.72}$$

resulting in the following set of equations:

$$\begin{cases} \dot{q}_j(\tau) = p_j(\tau), \\ \dot{p}_j(\tau) = -M_{ij}^{-1} K_{ij}\, q_j(\tau) - M_{ij}^{-1} C_{ij}\, p_j(\tau), \end{cases} \tag{7.73}$$

and in matrix form:

$$\begin{Bmatrix} \dot{q}_j \\ \dot{p}_j \end{Bmatrix} = \begin{bmatrix} \mathbf{0} & \mathbf{I} \\ -M_{ij}^{-1} K_{ij} & -M_{ij}^{-1} C_{ij} \end{bmatrix} \begin{Bmatrix} q_j \\ p_j \end{Bmatrix}, \tag{7.74}$$

where $\mathbf{0}$ is an $N \times N$ zero matrix and $\mathbf{I}$ is an $N \times N$ identity matrix. Equation (7.74) can be re-written as:

$$\dot{\mathbf{x}} = \mathrm{A}\mathrm{x}, \tag{7.75}$$

where

$$\mathbf{x} = \begin{Bmatrix} q_j \\ p_j \end{Bmatrix}, \tag{7.76}$$

and

$$\mathrm{A} = \begin{bmatrix} \mathbf{0} & \mathbf{I} \\ -M_{ij}^{-1}K_{ij} & -M_{ij}^{-1}C_{ij} \end{bmatrix}. \tag{7.77}$$

The general solution for the set of first-order ordinary differential equations of Eq. (7.75) is

$$\mathbf{x} = \mathbf{X}e^{\lambda t}. \tag{7.78}$$

By plugging this general form into Eq. (7.75), one finds:

$$\lambda \mathbf{X}e^{\lambda t} = \mathbf{A}\mathbf{X}e^{\lambda t}, \tag{7.79}$$

which returns the general form of an eigenvalue problem:

$$\lambda \mathbf{X} = \mathbf{A}\mathbf{X}. \tag{7.80}$$

The size of $\mathbf{X}$ is twice the number of modes used in the Galerkin discretization, i.e., $2N$. As we discussed in Chap. 5, when we solve an eigenvalue problem, we cannot find the magnitudes of the solution, but rather the eigenvalues of matrix $\mathbf{A}$. In this problem, matrix $\mathbf{A}$ depends on the system parameters, i.e., u and β. When u is varied, matrix $\mathbf{A}$ varies as well, and as a result the eigenvalues of matrix $\mathbf{A}$, and the stability of the system will depend on u. Therefore, to summarize the behavior of the pipe, we will use plots of the real part and imaginary parts of the eigenvalues versus u. We remember that the real part of the eigenvalue is proportional to a negative damping, and its imaginary part is proportional to the frequency of the system.

We first consider the behavior of the pipe when its both ends are supported (i.e., a simply supported or a clamped-clamped pipe), and then we will consider the case where one of the ends of the pipe is free, i.e., a cantilevered pipe. As we will see, the observed responses are completely different for these two sets of boundary conditions.

5 Linear Behavior of a Pipe with Supported Ends

Here, we consider the behavior of a pipe with supported boundaries. That would be a simply supported pipe (hinged-hinged boundary condition) or a clamped-clamped pipe. The similarity between these two boundary conditions is that at the boundaries, no movement is possible. We start with a simply supported boundary condition. To solve the pipe equation for this boundary condition, we need to use the mode shapes of a beam with simply supported boundary conditions:

$$\phi_j(\xi) = \sin(j\pi\xi), \tag{7.81}$$

in the Galerkin discretization. This means that first we calculate $A_{1,ij}$, $A_{2,ij}$, $A_{3,ij}$, and $A_{4,ij}$ (Eq. (7.65)) with these mode shapes, then calculate matrices **K**, **C**, and **M** (Eqs. (7.67) to (7.69)) for given values of u and β, and then calculate the eigenvalues of matrix **A** (Eq. (7.77)) to find a total of $2N$ eigenvalues, and plot their real parts and imaginary parts versus u. We repeat this process for each value of u, preferably starting at $u = 0$, where we know what to expect from the pipe (a pipe with zero flow velocity must be sitting still at its original equilibrium position) all the way to a maximum u of interest.

The plots of real and imaginary parts of the eigenvalue versus u are shown in Fig. 7.3 for a simply supported pipe. At $u = 0$, the real part of the eigenvalue corresponds to the damping of the structure, which is zero, because we have assumed that the structural damping is zero. The imaginary parts at $u = 0$ correspond to the natural frequencies of the structure, with no flow. We see that there are two natural frequencies (note that the negative values do not have any physical meaning, since there is no negative frequency), because in this discretization, we have used two structural modes, and each of these two natural frequencies corresponds to the natural frequency of one of those modes. As the flow velocity is increased, these frequencies change. The frequencies decrease with increasing u until the frequency of the first mode, the frequency that started around 10 at $u = 0$, reaches 0 at

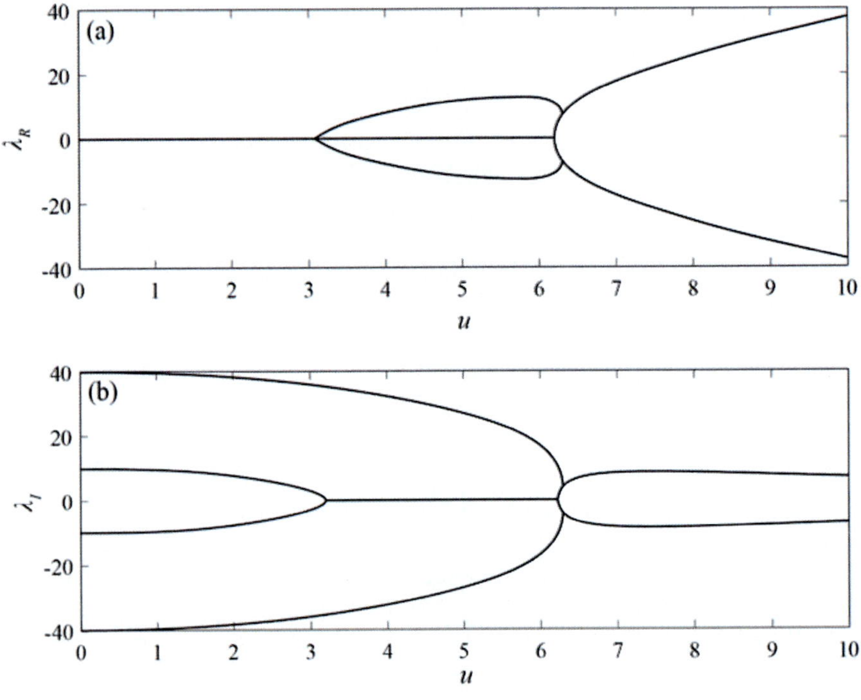

Fig. 7.3 (**a**) Real and (**b**) imaginary parts of the eigenvalues for a simply supported pipe conveying fluid

$u = 3.14 = \pi$. When the frequency of a structure goes to zero, the structure has no stiffness, and it buckles. Therefore, based on this plot, a simply supported pipe buckles at $u_b = \pi$. Starting at $u = \pi$, the real part of the eigenvalue becomes nonzero, and positive and negative dampings are observed. However, a negative damping, when there is no stiffness in the system, does not result in any dynamic response, and therefore is not physically significant. The second mode, whose frequency starts at ~40, follows a pattern similar to the first mode: it decreases with increasing flow velocity and it reaches the zero frequency at a dimensionless flow velocity of $u = 6.28 = 2\pi$, which implies a buckling of the second mode at this flow velocity. However, this second instability occurs well beyond the instability of the first mode at $u_b = \pi$, and we should keep in mind that we are solving a linear equation. The main assumption in deriving this linear equation was for the amplitude of the pipe to be small and to stay within the linear range. Once the pipe buckles, at $u_b = \pi$, the amplitude of its response could increase beyond the linear limit, where the model is no longer valid. Then, the stability of the second mode at a dimensionless flow velocity of $u = 2\pi$, which is predicted based on the linear model, is not reliable if the amplitude of the pipe's response at that flow velocity becomes very large. Therefore, and in general, we do not trust the instabilities that are predicted using a linear model, after the first point of instability, because they are based on a small-amplitude assumption, which could be a wrong assumption for instabilities observed after the first instability. The small-amplitude assumption is a fine assumption up to the first point of instability,[2] because the system, the pipe in our case, is at its original equilibrium position, about which the linear equation has been derived.

In summary, a pipe with a simply supported boundary condition buckles at $u_b = \pi$. Intuitively, we know that the pipe bends to a nonzero amplitude response when it buckles. However, we cannot show this using a linear model. We need a nonlinear model to calculate the amplitude of response, and we will discuss this nonlinear model later, in Sect. 7. It is also important to note that the critical value for buckling is independent from β. Any pipe, as long as it has simply supported boundary conditions, buckles at $u_b = \pi$. The value of β does not influence the critical value for buckling, because β appears only in dynamic terms of the equation of motion, and buckling is a static instability.

Now we consider a pipe with clamped boundaries, as another example of a case where the two boundaries of the pipe cannot move. All we need to do is to calculate matrix **A** using the mode shapes of a clamped-clamped beam. The rest of the method of solution will be identical to what we did for the simply supported boundary condition. The mode shapes of a clamped-clamped beam are (Table 6.2):

$$\phi_j(\xi) = \sinh \beta_n x - \sin \beta_n x + \alpha_n(\cosh \beta_n x - \cos \beta_n x), \tag{7.82}$$

where

[2]This statement is true unless a subcritical instability is observed, in which case the system could exhibit a nonzero response even before the first point of instability predicted by the linear model.

$$\alpha_n = \frac{\sinh \beta_n L - \sin \beta_n L}{\cosh \beta_n L - \cos \beta_n L}, \tag{7.83}$$

and

$$\begin{aligned} \beta_1 L &= 4.730, \\ \beta_2 L &= 7.853, \\ \beta_3 L &= 10.996, \\ \beta_4 L &= 14.137, \\ \beta_5 L &= 17.279 \\ \beta_n L &= \frac{(2n+1)\pi}{2}, \text{for } n > 5, \end{aligned} \tag{7.84}$$

where for our pipe, $L = 1$.

Using these mode shapes, we calculate matrix **A**, for a given β and for each dimensionless flow velocity, u. Then we plot the real and imaginary parts of the eigenvalue versus u and get what we see in Fig. 7.4. It is immediately clear that these plots look qualitatively very similar to those for the case of a simply supported pipe.

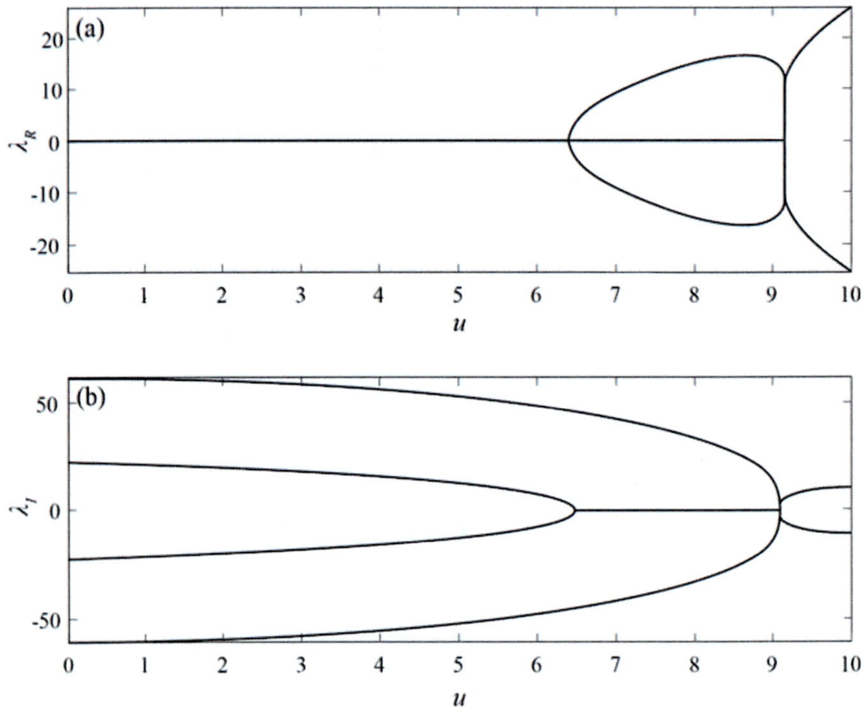

Fig. 7.4 (**a**) Real and (**b**) imaginary parts of the eigenvalues for a clamped-clamped pipe conveying fluid

The frequencies of the first two modes decrease with increasing u until the first mode's frequency becomes zero at $u_b = 6.28 = 2\pi$, resulting in buckling of the pipe in its first mode. Then a clamped-clamped pipe also buckles at a critical flow velocity as did the simply supported pipe; however, the critical flow velocity for buckling of a clamped-clamped pipe is twice that of a simply supported pipe. It is more difficult to buckle a clamped-clamped pipe than a simply supported pipe, which is intuitive. Similar to a simply supported pipe, after the pipe buckles, it bends to a nonzero amplitude; however, we cannot show this using a linear model. Also similar to the simply supported case, the critical value for the onset of buckling does not change with β, because β appears in dynamic terms only, and buckling is a static instability.

6 Linear Behavior of a Cantilevered Pipe

In this section, we study the behavior of a cantilevered pipe, as an example of a case where one of the ends of the pipe is free to move. Again, the only change that we have to make in our solution is to use the mode shapes of a cantilevered beam in calculating matrix **A**. These mode shapes are (Table 6.2):

$$\phi_n(x) = \cosh \beta_n x - \cos \beta_n x - \sigma_n(\sinh \beta_n x - \sin \beta_n x), \tag{7.85}$$

where

$$\sigma_n = \frac{\cos \beta_n L + \cosh \beta_n L}{\sin \beta_n L + \sinh \beta_n L}, \tag{7.86}$$

and

$$\begin{aligned}
\beta_1 L &= 1.875, \\
\beta_2 L &= 4.694, \\
\beta_3 L &= 7.855, \\
\beta_4 L &= 10.996, \\
\beta_5 L &= 14.137, \\
\beta_n L &= \frac{(2n-1)\pi}{2}, \text{for } n > 5.
\end{aligned} \tag{7.87}$$

We plug these mode shapes into the equation for matrix **A**, and calculate its eigenvalues for a given β and a range of dimensionless flow velocities, u. A plot of how the real and imaginary parts of the eigenvalues change with u is given in Fig. 7.5, for $\beta = 0.2$, when two modes are used in the Galerkin discretization. It is immediately clear that the general behavior is different from the previous two cases where the ends of the pipe were fixed. From the plots, it is clear that the frequency of

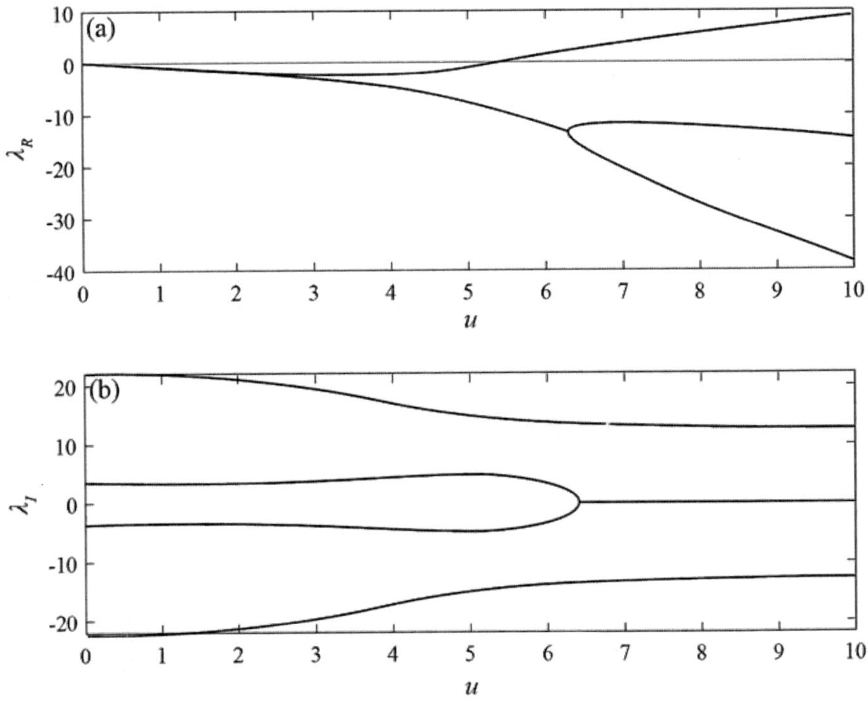

Fig. 7.5 (**a**) Real and (**b**) imaginary parts of the eigenvalues versus u for a cantilevered pipe conveying fluid for $\beta = 0.2$

the first mode increases initially and then decreases, and around $u = 6.4$, it becomes zero. However, the real parts of the eigenvalues act very differently from the previous cases. The real parts, which are proportional to a negative damping, start at zero, because there is no structural damping. Then they decrease (become negative), which means that the overall damping of the system is positive, i.e., if an initial disturbance is given to the pipe, it will damp out to zero after some cycles of oscillations—the damping behavior that we expect. Then the damping of one of the modes (it is not clear from this plot which mode it is—we will get back to this a bit later) starts increasing, and becomes positive at $u_f = 5.4$. The positive values of the real part that appear for $u > 5.4$ mean that the system has a negative damping: an initial disturbance of the system results in oscillations whose amplitude will increase with time, instead of decrease. Note that both modes of the structure have nonzero frequencies at $u_f = 5.4$, which means that oscillations are possible at this flow velocity. Then the system starts oscillating at $u_f = 5.4$, and according to the linear theory, the amplitude of these oscillations keeps increasing forever, because the system has a negative damping. Physically, we cannot observe oscillations whose amplitudes increase forever. However, a linear model cannot predict the amplitude of the response, and is not valid for large deviations from the original equilibrium anyway, and therefore we do not expect to get any meaningful amplitude information from it. The onset of oscillations, however, can be predicted based on the linear theory: A cantilevered pipe starts oscillating at a critical flow velocity. The observed oscillations are called *flutter*, and they start when the damping of the system becomes negative (a positive real part of the eigenvalue), while its frequency is nonzero.

The question that the plot of Fig. 7.5 cannot answer immediately is to which mode does the real part that crosses the zero line at the point of instability belong. Technically the eigenvalues come in a pair of real and imaginary numbers, but when they are all plotted versus u, it is hard to recognize them. If, however, we plot the imaginary part versus the real part, as is done for this example in Fig. 7.6, then we see which mode is the mode that becomes unstable. In the plot of Fig. 7.6, the behavior of the pair of eigenvalues is clear for both modes. The first mode is the mode with a smaller natural frequency, and clearly that would be the mode that starts at a lower point on the y-axis (the imaginary part, which is proportional to the frequency), and the second mode starts a bit higher. Then the real part of the first mode goes to the left, and stays on the negative side of the x-axis (the real part of the eigenvalue) and is always stable (positive damping). It is the second mode that initially goes to the left, with positive damping, and then turns and crosses the y-axis and goes to the positive side of the real part, which corresponds to a negative damping, and becomes unstable. At which flow velocity this occurs is not readily clear from the plot of Fig. 7.6, because the flow velocities are given implicitly. But it is clear which mode has become unstable. Then from the plot of Fig. 7.5 we know at what flow velocity the system becomes unstable, and from the plot of Fig. 7.6 we know which mode becomes unstable. The plot of Fig. 7.6 is called an Argand diagram.

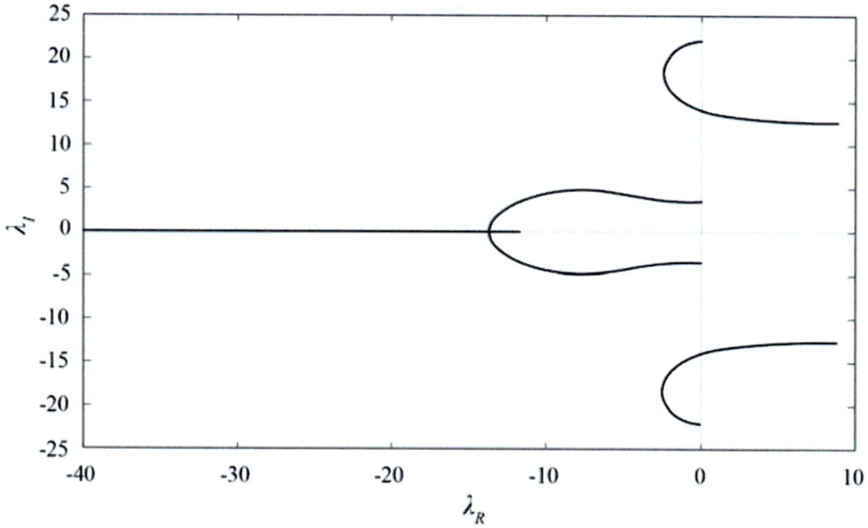

Fig. 7.6 An Argand diagram for a cantilevered pipe conveying fluid for $\beta = 0.2$

Then, from the plots of Figs. 7.5 and 7.6, it is clear that a cantilevered pipe conveying fluid is stable initially, and as the flow velocity is increased, at a critical flow velocity, the pipe undergoes a dynamic instability and starts oscillating. This dynamic instability is called a Hopf bifurcation. The critical value of u depends on β, because β shows up in dynamic terms of the equation of motion, and Hopf bifurcation is a dynamic instability. If we solve the linear equation of motion for a cantilevered pipe with different values of β, we can show that as β increases, the onset of instability increases.

We have seen in this section and in the previous section that a linear model can predict the onset of instability: after a critical flow velocity, a pipe supported at both ends buckles, and a cantilevered pipe flutters. How does the pipe look after these points of instabilities? If we are familiar with the concept of buckling from the past, we know that a buckled structure goes from its original stable position to another stable position: from one static solution (the trivial solution) to another static solution (the buckled position). What is the amplitude of a buckled pipe after its instability? Or for a cantilevered pipe that undergoes flutter, we know that there will be oscillations after the point of instability. What is the amplitude of these oscillations? Does the amplitude change with increasing flow velocity? Does the frequency change with increasing flow velocity? Is there any other instability after the first observed instability or will a buckled pipe stay buckled forever, and will a fluttering pipe keep fluttering forever? To answer these questions, or any other question about the system's behavior after the first point of instability, we should use a nonlinear model.

7 Holmes Nonlinear Equation for a Pipe with Supported Ends

To start discussing nonlinear models for a pipe conveying fluid, we first consider a model presented by Holmes (1978) for a pipe supported at both ends. From a linear analysis, we know that a pipe supported at both ends buckles at a critical flow velocity. With Holmes model for such pipes, we will be able to quantify what occurs after the instability, specifically, we will observe how the amplitude of a buckled beam changes with increasing flow velocity beyond the point of instability.

Holmes nonlinear model is perhaps the simplest possible nonlinear model for a pipe. Holmes only added one nonlinear term to the linear equation of motion to create a nonlinear model. This single term represents the axial tension that is induced in the pipe as it buckles in the lateral direction, while fixed at its both ends. This tension can be written as:

$$T = \sigma A, \tag{7.88}$$

in which σ is the stress in the pipe and A is its cross-section. Then stress can be related to strain, ε, and the relation for the tension can be re-written as:

$$T = E\varepsilon A, \tag{7.89}$$

in which ε is the averaged axial strain, and can be related to the slope of the beam, y', as

$$\varepsilon = \frac{1}{2L}\int_0^L y'^2 dx, \tag{7.90}$$

and therefore, the tension becomes:

$$T = \frac{EA}{2L}\int_0^L y'^2 dx. \tag{7.91}$$

This is the term that Holmes added to the linear equation of motion of the pipe, as an added nonlinear tension. This is clearly a nonlinear term, because of the appearance of y'^2 in it, and since this is a tension term, its sign must be opposite the sign of the flexural rigidity term, as we observed in the general equation of motion for a beam in Chap. 6:

$$EIy'''' - Ty'' + m\ddot{y} = 0. \tag{7.92}$$

Now we consider the linear equation for a pipe:

$$EIy'''' + MU^2y'' + 2MU\dot{y}' + (m+M)\ddot{y} = 0, \tag{7.93}$$

and add the nonlinear tension term to it, to get Holmes nonlinear equation for a pipe supported at both ends:

$$EIy'''' - \left(\frac{EA}{2L}\int_0^L y'^2 dx\right)y'' + MU^2y'' + 2MU\dot{y}' + (m+M)\ddot{y} = 0. \tag{7.94}$$

We see that the nonlinear tension term is "competing" with the linear compression term, MU^2y''. It is the compression term that causes buckling, and it is the new nonlinear tension term that keeps the pipe's amplitude from going to infinity after buckling.

Before anything else, we make Eq. (7.94) dimensionless, the same way that we made the linear equation of motion dimensionless. In the process of nondimensionalization, we introduce a new dimensionless parameter:

$$\Pi_0 = \frac{EAL^2}{EI}, \tag{7.95}$$

which relates the flexibility of the pipe in the axial direction, EAL^2, to the flexibility of the pipe in the transverse direction, *EI*. The dimensionless form of Holmes equation then becomes:

$$\eta'''' + \left(u^2 - \frac{1}{2}\Pi_0 \int_0^1 \eta'^2 d\xi\right)\eta'' + 2\beta^{\frac{1}{2}} u\dot{\eta}' + \ddot{\eta} = 0. \tag{7.96}$$

It ought to be mentioned here that Holmes original equation also included two additional structural damping terms, which we do not have in Eq. (7.96). Clearly, the structure's damping does not change the static response of a buckling pipe—damping is a dynamic term, and buckling is a static response. However, it might be necessary to use some structural damping in the system for the numerical solutions to converge in some numerical methods. Here, we add a viscous damping term to the system as $2\zeta\dot{\eta}$:

$$\eta'''' + \left(u^2 - \frac{1}{2}\Pi_0 \int_0^1 \eta'^2 d\xi\right)\eta'' + 2\beta^{\frac{1}{2}} u\dot{\eta}' + 2\zeta\dot{\eta} + \ddot{\eta} = 0, \tag{7.97}$$

where ζ is a damping ratio.

8 Solving the Holmes Equation

To solve the Holmes equation, we use the Galerkin technique. This time, after discretizing the equation of motion using the Galerkin technique, we will find a set of nonlinear ordinary differential equations (ODEs), instead of an eigenvalue problem that we found in the case of a linear system. By solving the set of nonlinear ODEs, we will find the magnitude and frequency of response. To apply the Galerkin technique, we assume the solution to be:

$$\eta(\xi, \tau) = \sum_{j=1}^{N} \phi_j(\xi)\, q_j(\tau), \tag{7.98}$$

in which $\phi_j(\xi)$ are the mode shapes of a beam supported at both ends (e.g., a simply supported or clamped-clamped beam). Then we plug $\eta(\xi, \tau)$ of Eq. (7.98) into the nonlinear equation of motion (Eq. (7.97)) to get:

$$\sum_{j=1}^{N} \phi_j''''(\xi)\, q_j(\tau) + \left(u^2 - \frac{1}{2}\Pi_0 \int_0^1 \sum_{k=1}^{N} \phi_k'(\xi) \sum_{l=1}^{N} \phi_l'(\xi)\, q_k(\tau) q_l(\tau) d\xi\right) \sum_{j=1}^{N} \phi_j''(\xi)\, q_j(\tau)$$
$$+ 2\beta^{\frac{1}{2}} u \sum_{j=1}^{N} \phi_j'(\xi)\, \dot{q}_j(\tau) + 2\zeta \sum_{j=1}^{N} \phi_j(\xi)\, \dot{q}_j(\tau) + \sum_{j=1}^{N} \phi_j(\xi)\, \ddot{q}_j(\tau) = 0, \tag{7.99}$$

and multiply everything by $\phi_i(\xi)$ and integrate over the entire length to get:

$$\begin{aligned}
&\sum_{j=1}^{N} \int_0^1 \phi_i(\xi)\phi_j''''(\xi)\mathrm{d}\xi\, q_j(\tau) + u^2 \sum_{j=1}^{N} \int_0^1 \phi_i(\xi)\phi_j''(\xi)d\xi\, q_j(\tau) \\
&-\frac{1}{2}\Pi_0 \sum_{j=1}^{N}\sum_{k=1}^{N}\sum_{l=1}^{N} \int_0^1 \left(\phi_i(\xi)\phi_j''(\xi)\int_0^1 \phi_k'(\xi)\phi_l'(\xi)d\xi\right) d\xi\, q_j(\tau)q_k(\tau)q_l(\tau) \\
&+ 2\beta^{\frac{1}{2}}u \sum_{j=1}^{N} \int_0^1 \phi_i(\xi)\phi_j'(\xi)\mathrm{d}\xi\, \dot{q}_j(\tau) + 2\zeta \sum_{j=1}^{N} \int_0^1 \phi_i(\xi)\phi_j(\xi)\,\mathrm{d}\xi\, \dot{q}_j(\tau) \\
&+ \sum_{j=1}^{N} \int_0^1 \phi_i(\xi)\phi_j(\xi)\mathrm{d}\xi\, \ddot{q}_j(\tau) = 0.
\end{aligned} \tag{7.100}$$

Then we define:

$$\begin{aligned}
A_{ij} &= \int_0^1 \phi_i(\xi)\phi_j''''(\xi)\mathrm{d}\xi, \\
B_{ij} &= \int_0^1 \phi_i(\xi)\phi_j''(\xi)d\xi, \\
C_{ijkl} &= \int_0^1 (\phi_i(\xi)\phi_j''(\xi)\int_0^1 \phi_k'(\xi)\phi_l'(\xi)d\xi)d\xi, \\
D_{ij} &= \int_0^1 \phi_i(\xi)\phi_j'(\xi)\mathrm{d}\xi, \\
E_{ij} &= \int_0^1 \phi_i(\xi)\phi_j(\xi)\mathrm{d}\xi,
\end{aligned} \tag{7.101}$$

and write the equations of motion as:

$$\begin{aligned}
&\sum_{j=1}^{N} A_{ij}q_j(\tau) + u^2 \sum_{j=1}^{N} B_{ij}\, q_j(\tau) - \frac{1}{2}\Pi_0 \sum_{j=1}^{N}\sum_{k=1}^{N}\sum_{l=1}^{N} C_{ijkl}\, q_j(\tau)q_k(\tau)q_l(\tau) \\
&+2\beta^{\frac{1}{2}}u \sum_{j=1}^{N} D_{ij}\dot{q}_j(\tau) + 2\zeta \sum_{j=1}^{N} E_{ij}\, \dot{q}_j(\tau) + \sum_{j=1}^{N} E_{ij}\, \ddot{q}_j(\tau) = 0,
\end{aligned} \tag{7.102}$$

and by collecting the stiffness, damping, inertia, and nonlinear terms together, we write the equations as:

$$\begin{aligned}
&\left(\sum_{j=1}^{N} A_{ij} + u^2 \sum_{j=1}^{N} B_{ij}\right) q_j(\tau) + \left(2\beta^{\frac{1}{2}}u \sum_{j=1}^{N} D_{ij} + 2\zeta \sum_{j=1}^{N} E_{ij}\right)\dot{q}_j(\tau) + \sum_{j=1}^{N} E_{ij}\, \ddot{q}_j(\tau) \\
&-\frac{1}{2}\Pi_0 \sum_{j=1}^{N}\sum_{k=1}^{N}\sum_{l=1}^{N} C_{ijkl}\, q_j(\tau)q_k(\tau)q_l(\tau) = 0,
\end{aligned} \tag{7.103}$$

or in the matrix form as:

$$\mathbf{K}q_j + \mathbf{C}\dot{q}_j + \mathbf{M}\ddot{q}_j - \mathbf{N}\, q_jq_kq_l = 0. \tag{7.104}$$

This matrix form of equations clearly shows that the equations of motion are combinations of all the linear terms that we had seen in the linear model of the system, plus one nonlinear term, $\mathbf{N}\, q_jq_kq_l$. The nonlinear term is of the third order, a cubic term, and consists of multiplications of three displacement quantities: q_j, q_k, and q_l. This is then a cubic stiffness term, a static term that can give us information on what a buckled pipe does.

To solve the set of ODEs of Eq. (7.104), we first write the equations in the first-order form by assuming $\dot{q}_j = p_j$, and plugging this assumption into Eq. (7.104):

$$\begin{aligned} \dot{q}_j &= p_j, \\ \dot{p}_j &= -\mathbf{M}^{-1}\mathbf{K}q_j - \mathbf{M}^{-1}\mathbf{C}p_j + \mathbf{M}^{-1}\mathbf{N}\, q_j q_k q_l. \end{aligned} \tag{7.105}$$

Then by defining a vector:

$$\mathbf{r} = \begin{Bmatrix} \mathbf{q} \\ \mathbf{p} \end{Bmatrix}, \tag{7.106}$$

the equations of motion can be written as:

$$\dot{\mathbf{r}} = \begin{Bmatrix} \dot{\mathbf{q}} \\ \dot{\mathbf{p}} \end{Bmatrix} = \begin{bmatrix} \mathbf{0} & \mathbf{I} \\ -\mathbf{M}^{-1}\mathbf{K} + \mathbf{M}^{-1}\mathbf{N}\, q_k q_l & -\mathbf{M}^{-1}\mathbf{C} \end{bmatrix} \begin{Bmatrix} \mathbf{q} \\ \mathbf{p} \end{Bmatrix}. \tag{7.107}$$

Note that the size of the matrix in Eq. (7.107) is $2N$, as after using N modes to discretize the original PDE, we end up getting a set of N second-order ODEs (Eq. (7.104)), which we write in terms of a set of $2N$ first-order ODEs. Also, note that the matrix in Eq. (7.107) contains the unknowns q_k and q_l, which make the set of equations in Eq. (7.107) a set of nonlinear equations of motion. We can now use any nonlinear ODE solver and solve Eq. (7.107) and obtain q_j, our unknowns. After finding all q_j, we will go back to our original assumption:

$$\eta(\xi, \tau) = \sum_{j=1}^{N} \phi_j(\xi)\, q_j(\tau), \tag{7.108}$$

and find $\eta(\xi, \tau)$, the dimensionless displacement of the pipe in the transverse direction as it varies along the length of the pipe and with time. It is this quantity that we will plot in the figures that summarize the response of the structure.

9 The Post-critical Behavior of a Pipe Supported at Both Ends

For low flow velocities (flow velocities smaller than the critical value for buckling), the pipe stays at its original equilibrium position. To solve the equations of motion, we need to provide a set of nonzero initial conditions. Physically, this is equivalent of disturbing the pipe at time zero by, for example, giving an initial displacement to it, and then letting it go. The pipe then oscillates, and the amplitude of oscillations decreases with time and approaches its original equilibrium—zero amplitude. This is

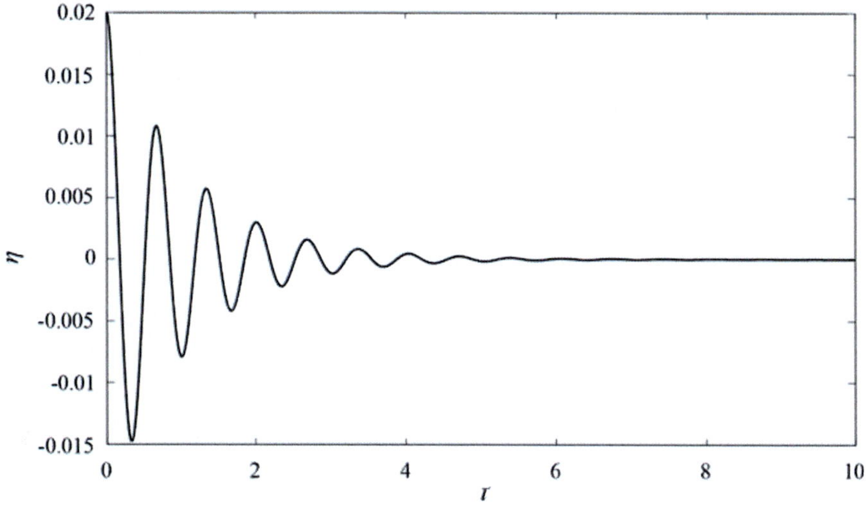

Fig. 7.7 The response of a simply supported pipe with $\beta = 0.2$, $\Pi_0 = 1000$, $\zeta = 0.95$, and $u = 1$ using two modes in the Galerkin discretization

shown for a sample case of low flow velocity in Fig. 7.7. Oscillations start at a nonzero initial value (the given initial condition), and approach zero after some cycles. The steady-state solution is therefore zero—the original position of the pipe with no flow velocity.

At higher flow velocities (i.e., flow velocities larger than the critical), instead of approaching zero, the solution approaches a nonzero, but constant value, as shown in the sample case of Fig. 7.8. This nonzero constant value represents the amplitude of the midpoint of the pipe at a given flow velocity ($u = 5$ in this example), and it shows that the pipe's stable solution is at a finite nonzero value. This value of flow velocity is larger than the critical value for buckling that we had found using our linear analysis ($u_b = \pi$). We had predicted using the linear analysis that the pipe buckles at this critical value. Now, using a nonlinear model, we can find the magnitude of response after the instability.

To show how the pipe's amplitude changes with increasing flow velocity, we plot the magnitude of the steady-state response of the midpoint of the pipe versus the flow velocity in Fig. 7.9. This plot is sometimes also called a bifurcation diagram, because it summarizes the instabilities that the system undergoes. Figure 7.9 shows that, in agreement with the linear results, the pipe stays stable up to $u_b = \pi$, where it undergoes a pitchfork bifurcation and it buckles. After buckling, the magnitude of the static deflection of the pipe increases with increasing flow velocity. We show the results up to $u = 8$ in this plot, but for even higher flow velocities, the same pattern is observed: the amplitude keeps increasing with increasing flow velocity.

We can use the results of our nonlinear model to also show how the pipe's amplitude changes along its length at any given flow velocity. Since using the linear model we predicted that the first mode of the pipe becomes unstable, we expect the pipe to follow a half-sine shape in its buckled position—the shape that represents

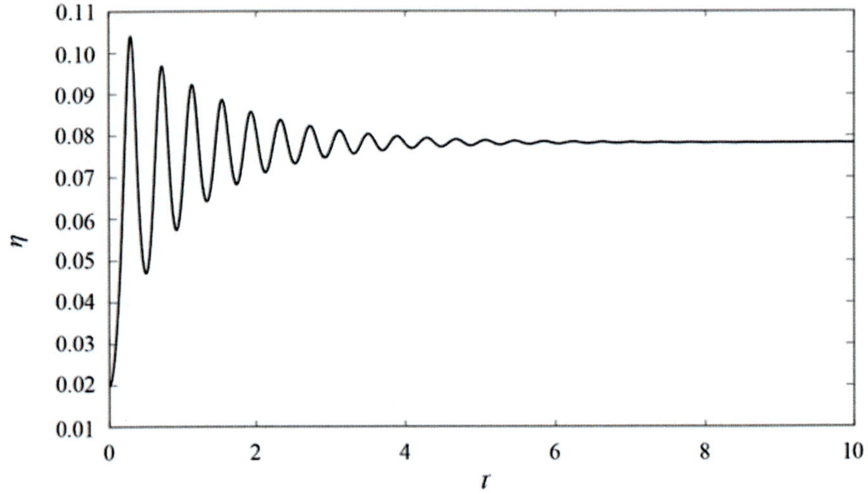

Fig. 7.8 The response of a simply supported pipe with $\beta = 0.2$, $\Pi_0 = 1000$, $\zeta = 0.95$, and $u = 5$ using two modes in the Galerkin discretization

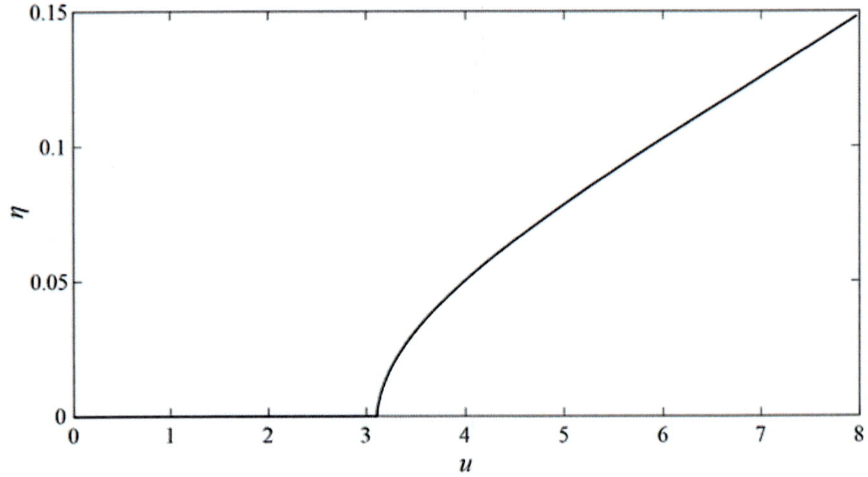

Fig. 7.9 A bifurcation diagram for a simply supported pipe conveying fluid with $\beta = 0.2$ and $\Pi_0 = 1000$, using two modes in the Galerkin discretization

the first mode of a simply supported beam. This is in fact what we observe if we plot the amplitude of the pipe along its length, as shown for $u = 5$ in Fig. 7.10.

The nonlinear results for a pipe with clamped-clamped boundary conditions will be very similar to those of a simply supported pipe. There will be a critical point for buckling at $u_{cr} = 2\pi$ as predicted by the linear model, and then the amplitude of the buckled pipe will increase with increasing flow velocity. The shape of the pipe will be slightly different to represent the first mode of a clamped-clamped beam, rather than a simply supported beam. Other than that, the qualitative behavior will be

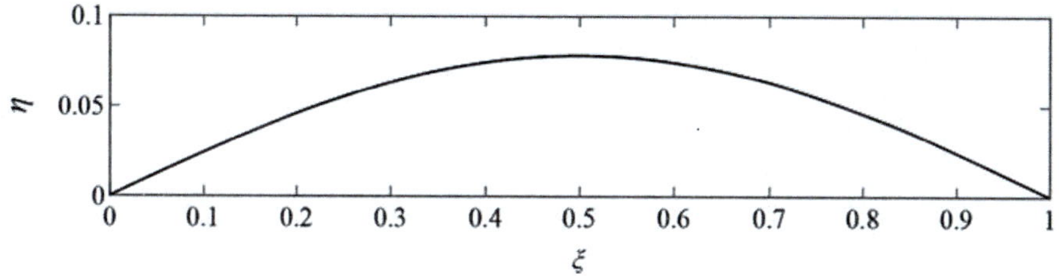

Fig. 7.10 The buckled state of a simply supported pipe with $\beta = 0.2$, $\Pi_0 = 1000$, $\zeta = 0.95$, and $u = 5$ using two modes in the Galerkin discretization

the same: increased amplitude of response with increasing flow velocity for post-critical range, with no secondary instabilities. There is no dynamic response in the post-buckling behavior of a pipe supported at both ends.

In the following section, we will consider a cantilevered pipe for which the linear analysis predicted a flutter instability as the first expected instability. We expect to observe limit cycle oscillations after this dynamic instability.

10 A Nonlinear Cantilevered Pipe

To study the nonlinear behavior of a cantilevered pipe, we need a different equation. The only nonlinear term in the Holmes equation is a tension term that is produced because the two ends of the structure are fixed. This is obviously not the case in a cantilevered pipe, since one of the two ends is free. Here, we consider a more general equation of motion for a cantilevered pipe, without going through its derivation. One of the simplest nonlinear equations that represents the motion of a cantilevered pipe is (Païdoussis, 2016):

$$\begin{aligned} &EI(y'''' + y''''y'^2 + 4y'y''y''' + y''^3) + (m+M)\ddot{y} + 2MU\dot{y}'(1+y'^2) + MU^2(y'' + y''y'^2) \\ &\quad -y''[\textstyle\int_x^L\int_0^x (m+M)(\dot{y}'^2 + y'\ddot{y}')dxdx + \int_x^L (2MUy'\dot{y}' + MU^2y'y'')dx] \\ &\quad +y'\textstyle\int_0^x (m+M)(\dot{y}'^2 + y'\ddot{y}')dx = 0. \end{aligned} \tag{7.109}$$

There are several nonlinear terms in this equation. Before we discuss what each of them represents, we should note that the linear terms that we had seen in the linear equation of motion of the pipe all exist in this nonlinear equation. Besides the linear terms, all the nonlinear terms that we observe in this equation are of order three—cubic nonlinearities. This is because in the derivation of the equation, which we did not discuss here, every time the Taylor expansion is used to expand a nonlinear term, terms up to the third order are kept, instead of only the first order, which we did when we derived the linear equation of motion. This is called a weakly nonlinear model, since the nonlinear terms have been obtained after Taylor expansions.

The first four terms in the equation of motion, $EI(y'''' + y''''y'^2 + 4y'y''y''' + y''^3)$, represent the flexural rigidity of the beam. The first term of these four is the linear flexural rigidity, and the other three are the nonlinear terms. These terms are also seen in the nonlinear Euler-Bernoulli beam equation, which is a nonlinear representation of a beam. The next term, $(m + M)\ddot{y}$, is the linear inertia term, and we have seen it also in the linear equation of motion. This term is then followed by the linear and nonlinear Coriolis force terms, $2MU\dot{y}'(1 + y'^2)$, and centripetal terms: $MU^2(y'' + y''y'^2)$. The terms in the second line of Eq. (7.109) act as a nonlinear tension. Note that these terms all have a $-y''$ in common, and their sign is opposite the sign of the flexural rigidity term, and therefore act as nonlinear tension terms. It is important to note that these nonlinear tension terms are functions of the pipe's velocity as well, and therefore do not influence only the static response of the system, but also its possible dynamic response. The last terms in the equation of motion (the terms in the third line of Eq. (7.109)) are nonlinear damping and inertia terms.

The equation of motion can be nondimensionalized using the same two dimensionless parameters that we had already seen in the linear equation, β and u, to get:

$$\begin{aligned} &\eta'''' + \eta''''\eta'^2 + 4\eta'\eta''\eta''' + \eta''^3 + \ddot{\eta} + 2u\sqrt{\beta}\dot{\eta}'(1 + \eta'^2) + u^2(\eta'' + \eta''\eta'^2) \\ &\quad -\eta''\left[\int_\xi^1 \int_0^\xi (\dot{\eta}'^2 + \eta'\ddot{\eta}')d\xi d\xi + \int_\xi^1 (2u\sqrt{\beta}\eta'\dot{\eta}' + u^2\eta'\eta'')d\xi\right] \\ &\quad +\eta' \int_0^\xi (\dot{\eta}'^2 + \eta'\ddot{\eta}')d\xi = 0. \end{aligned} \tag{7.110}$$

Note that Π_0, which was the coefficient of the only nonlinear term in the Holmes equation, does not exist here. To solve this nonlinear partial differential equation, we use the Galerkin technique, the same way we used it to solve the Holmes equation. Here, we have more nonlinear terms, and as a result there will be more terms in the discretized version of the equation. After assuming:

$$\eta(\xi, \tau) = \sum_{j=1}^{N} \phi_j(\xi)\, q_j(\tau), \tag{7.111}$$

and plugging it into the equation of motion, multiplying everything by $\phi_i(\xi)$, integrating from 0 to 1, and collecting all similar terms, we get:

$$\begin{aligned} &(A_{1,ij} + u^2 A_{2,ij})q_j + 2u\sqrt{\beta}A_{3,ij}\dot{q}_j + A_{4,ij}\ddot{q}_j \\ &+ 2u\sqrt{\beta}(B_{1,ijkl} - B_{6,ijkl})q_j q_k \dot{q}_l + [u^2(B_{2,ijkl} - B_{7,ijkl}) + B_{3,ijkl} + 4B_{4,ijkl} + B_{5,ijkl}]q_j q_k q_l \\ &- (B_{8,ijkl} + B_{9,ijkl})q_j \dot{q}_k \dot{q}_l - (B_{10,ijkl} + B_{11,ijkl})\ddot{q}_j q_k q_l = 0, \end{aligned} \tag{7.112}$$

in which

$$
\begin{aligned}
A_{1,ij} &= \int_0^1 \phi_i \phi_j'''' \, \mathrm{d}\xi, \\
A_{2,ij} &= \int_0^1 \phi_i \phi_j'' \mathrm{d}\xi, \\
A_{3,ij} &= \int_0^1 \phi_i \phi'_j \mathrm{d}\xi, \\
A_{4,ij} &= \int_0^1 \phi_i \phi_j \mathrm{d}\xi, \\
B_{1,ijkl} &= \int_0^1 \phi_i \phi'_j \phi'_k \phi'_l \mathrm{d}\xi, \\
B_{2,ijkl} &= \int_0^1 \phi_i \phi''_j \phi'_k \phi'_l \mathrm{d}\xi, \\
B_{3,ijkl} &= \int_0^1 \phi_i \phi''''_j \, \phi'_k \phi'_l \mathrm{d}\xi, \\
B_{4,ijkl} &= \int_0^1 \phi_i \phi'_j \phi''_k \phi'''_l \, \mathrm{d}\xi, \\
B_{5,ijkl} &= \int_0^1 \phi_i \phi''_j \phi''_k \phi''_l \, \mathrm{d}\xi, \\
B_{6,ijkl} &= \int_0^1 \phi_i \phi''_l \int_\xi^1 \phi'_k \phi'_j \mathrm{d}\xi \mathrm{d}\xi, \\
B_{7,ijkl} &= \int_0^1 \phi_i \phi''_j \int_\xi^1 \phi'_k \phi''_l \mathrm{d}\xi \mathrm{d}\xi, \\
B_{8,ijkl} &= \int_0^1 \phi_i \phi''_j \int_\xi^1 \int_0^\xi \phi_k \phi_l \mathrm{d}\xi \mathrm{d}\xi \mathrm{d}\xi, \\
B_{9,ijkl} &= \int_0^1 \phi_i \phi'_j \int_\xi^1 \phi'_k \phi'_l \mathrm{d}\xi \mathrm{d}\xi, \\
B_{10,ijkl} &= \int_0^1 \phi_i \phi''_l \int_\xi^1 \int_0^\xi \phi'_k \phi_j \mathrm{d}\xi \mathrm{d}\xi \mathrm{d}\xi, \\
B_{11,ijkl} &= \int_0^1 \phi_i \phi'_l \int_\xi^1 \phi'_k \phi'_j \mathrm{d}\xi \mathrm{d}\xi
\end{aligned}
\tag{7.113}
$$

Here, all the A terms are the linear terms that also existed in the linear equation of motion, and all the B terms are the nonlinear terms. Equation (7.112) then can be written as:

$$
\begin{aligned}
&K_{ij}q_j + C_{ij}\dot{q}_j + M_{ij}\ddot{q}_j \\
&+KN_{ij}q_jq_kq_l + CN_{1,ij}q_jq_k\dot{q}_l - CN_{2,ij}q_j\dot{q}_k\dot{q}_l - MN_{ij}\ddot{q}_jq_kq_l = 0,
\end{aligned}
\tag{7.114}
$$

in which

$$
\begin{aligned}
K_{ij} &= A_{1,ij} + u^2A_{2,ij}, \\
C_{ij} &= 2u\sqrt{\beta}A_{3,ij}, \\
M_{ij} &= A_{4,ij}, \\
CN_{1,ij} &= 2u\sqrt{\beta}\left(B_{1,ijkl} - B_{6,ijkl}\right), \\
CN_{2,ij} &= B_{8,ijkl} + B_{9,ijkl}, \\
MN_{ij} &= B_{10,ijkl} + B_{11,ijkl}, \\
KN_{ij} &= u^2\left(B_{2,ijkl} - B_{7,ijkl}\right) + B_{3,ijkl} + 4B_{4,ijkl} + B_{5,ijkl}.
\end{aligned}
\tag{7.115}
$$

Eq. (7.114) includes both linear and nonlinear terms. The linear terms, $K_{ij}q_j$, $C_{ij}\dot{q}_j$ and $M_{ij}\ddot{q}_j$ are the linear stiffness, damping, and inertia terms. The nonlinear term $KN_{ij}q_jq_kq_l$ represents a nonlinear cubic stiffness. Two nonlinear terms, $CN_{1,ij}q_jq_k\dot{q}_l$ and $CN_{2,ij}q_j\dot{q}_k\dot{q}_l$, represent nonlinear damping, and the nonlinear term $MN_{ij}\ddot{q}_jq_kq_l$ represents nonlinear inertia. It is important to note that the two terms which include both velocity and displacement could be considered as either nonlinear stiffness terms or nonlinear damping terms, and the nonlinear inertia term could technically be considered a nonlinear stiffness term, because it also contains q_kq_l.

11 Typical Nonlinear Behavior of a Cantilevered Pipe

We can solve Eq. (7.114), using any numerical method that is suitable for solving a set of nonlinear ordinary differential equations, and find the contributions of each mode in the total response, $q_j(\tau)$. Then we use the original assumption in the Galerkin technique, (Eq. (7.111)) to find the response along the length of the pipe versus time. For small values of u, an initial disturbance of the pipe is damped and the solution converges to zero: the original equilibrium position of the pipe—in agreement with linear predictions. As u is increased beyond a critical value, an initial disturbance of the pipe is not damped anymore, but instead leads to steady-state limit cycle oscillations. This is expected based on the linear results, which predicted a Hopf bifurcation to occur for this system. The response that is observed after a Hopf bifurcation occurs is a limit cycle. A sample of the limit cycle oscillations is shown in Fig. 7.11 for a pipe with $\beta = 0.2$ at $u = 6$.

Oscillations along the length of the pipe can also be observed. Several snapshots of the pipe's oscillations at different times as it oscillates are shown in Fig. 7.12. From the snapshots, it becomes clear how the pipe oscillates during this limit cycle

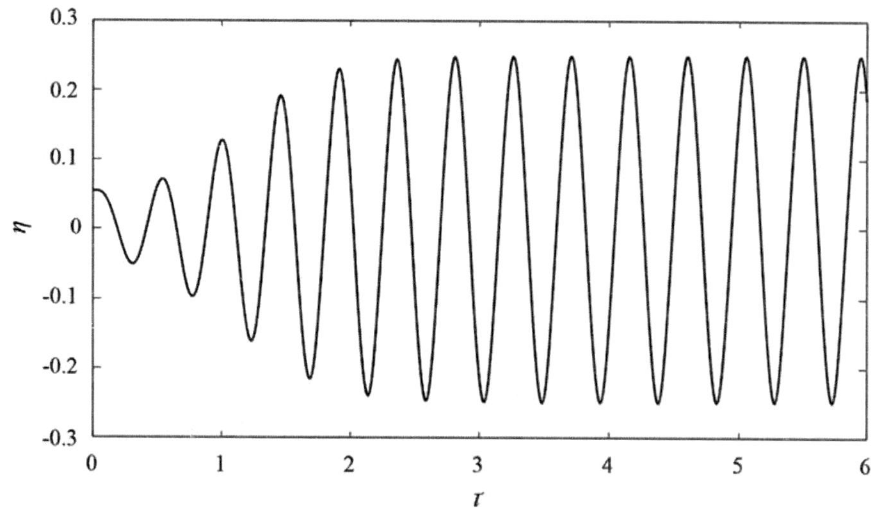

Fig. 7.11 Time history of the response of the tip of a cantilevered pipe for $\beta = 0.2$ at $u = 6$

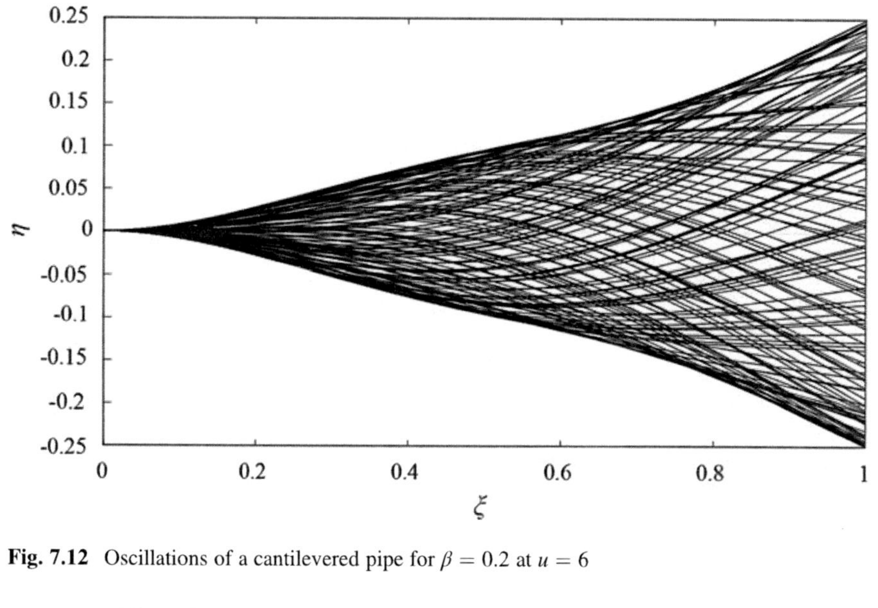

Fig. 7.12 Oscillations of a cantilevered pipe for $\beta = 0.2$ at $u = 6$

response. The pipe's shape resembles that of the second mode shape of a cantilevered beam, as a node is observed along its length. These limit cycle oscillations are the result of the second mode of the pipe becoming unstable, as the linear results predicted, and we can observe this fact in the snapshots of the pipe's oscillations as well.

To summarize the response of a cantilevered pipe, we plot the magnitude of the response versus the dimensionless velocity in a bifurcation diagram. This bifurcation

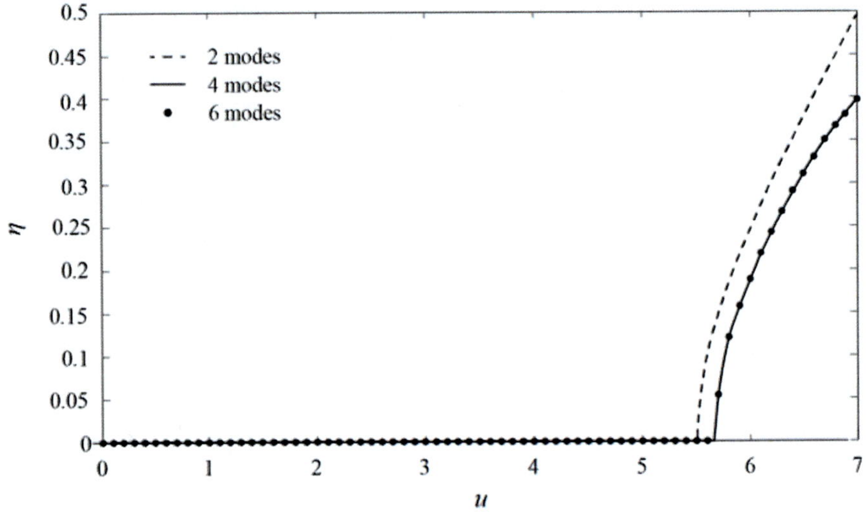

Fig. 7.13 A bifurcation diagram for the response of a cantilevered pipe conveying fluid for $\beta = 0.2$ using 2 modes (dashed line), 4 modes (continuous line), and 6 modes (black dots)

diagram, for a pipe with $\beta = 0.2$, is shown in Fig. 7.13. These amplitudes are the amplitudes of oscillations at each flow velocity. Since only the amplitude of oscillations is plotted, the bifurcation diagram might look like the bifurcation diagram of a system with a static response, but one should keep in mind that a frequency is also associated with each amplitude plotted in this bifurcations diagram, as the response is a limit cycle.

Three different versions of the bifurcation diagram are plotted in Fig. 7.13, based on the results obtained from different numbers of modes used in the Galerkin discretization. There are obvious differences between the results using only 2 modes, and the results using 4 or 6 modes. The bifurcation diagram of the results based on 2 modes shows an instability point at a critical value lower than the critical value predicted by the results based on 4 or 6 modes. The amplitude of oscillations is also larger at any given flow velocity when 2 modes are used. This reminds us of a very important part of using the Galerkin method in discretizing a PDE, and that is a convergence test. Clearly using only two modes in discretizing the equation of motion for a pipe conveying fluid is not enough to obtain reliable results, because when we use 4 or 6 modes, the results are different. Based on the bifurcation diagram of Fig. 7.13, four is the minimum number of modes we need in this discretization, because the results based on 4 modes and based on 6 modes are the same. It is important to note that one might think that since the pipe is excited at its second mode, then using only 2 modes in the discretization must be enough. This is not the case, because (i) the modes that we use in discretization are the mode shapes of a beam without any fluid inside, and therefore the second mode of a pipe with fluid inside is not necessarily the same, and (ii) when we solve the set of ODEs that results from the discretization of the PDE using the Galerkin method, we solve a set of coupled ODEs. Even if the contributions of higher modes in the final solution are

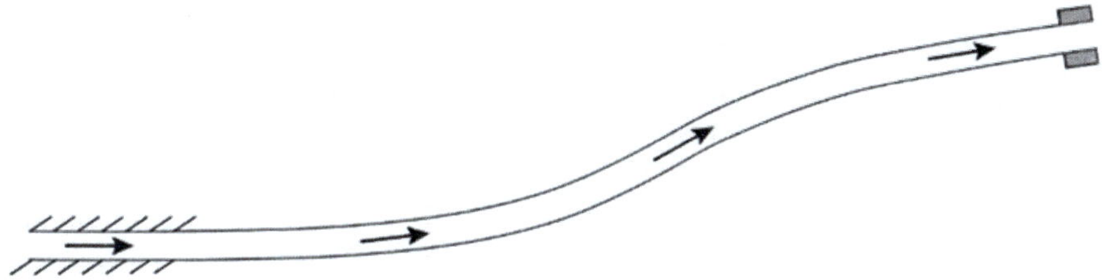

Fig. 7.14 A cantilevered pipe conveying fluid with an end mass

small, they could influence the contributions of the lower modes when a coupled set of equations is solved. When using the Galerkin method, one must always conduct a convergence test by increasing the number of modes used in the discretization until the difference in the obtained results becomes negligible.

12 A Cantilevered Pipe with an End Mass

After a cantilevered pipe becomes unstable through a Hopf bifurcation instability, it undergoes limit cycle oscillations. No other instability occurs for higher flow velocities. However, if the pipe is "modified" slightly, more complicated dynamic responses can be observed. These modifications can come in different forms, for example, by adding a mass or a spring somewhere along the length of the pipe, or by adding a nozzle at the tip of the pipe. Here we consider a case where a mass is added to the tip of the pipe, as an example. Firehoses have a nozzle at the tip, and the mass of the nozzle is usually non-negligible. This makes holding the firehose more difficult, especially if the flow velocity is high—which is normally the case for flow inside a firehose. By adding a mass at the tip of a cantilevered pipe, Copeland and Moon (1992) observed various routes to chaos in a pipe conveying fluid. Chaotic responses of this system are highly three-dimensional and in order to reproduce them we need three-dimensional models. Here, we first consider a two-dimensional (planar) case where an end mass is attached to the tip of the pipe.

The equation of motion in its dimensionless form will have the following form (Semler et al., 1994):

$$
\begin{aligned}
&\eta'''' + \eta''''\eta'^2 + 4\eta'\eta''\eta''' + \eta''^3 + \ddot{\eta} + 2u\sqrt{\beta}\dot{\eta}'(1+\eta'^2) + u^2(\eta'' + \eta''\eta'^2) \\
&\quad - \eta''\left[\int_\xi^1 \int_0^\xi (\dot{\eta}'^2 + \eta'\ddot{\eta}')d\xi d\xi + \int_\xi^1 (2u\sqrt{\beta}\eta'\dot{\eta}' + u^2\eta'\eta'')d\xi\right] \\
&\quad + \eta' \int_0^\xi (\dot{\eta}'^2 + \eta'\ddot{\eta}')d\xi + \mu\delta(\xi - 1)\ddot{\eta} \\
&\quad - \eta'' \int_\xi^1 \mu\delta(\xi - 1) \int_0^\xi (\dot{\eta}'^2 + \eta'\ddot{\eta}')d\xi d\xi \\
&\quad + \mu\delta(\xi - 1)\eta' \int_0^\xi (\dot{\eta}'^2 + \eta'\ddot{\eta}')d\xi = 0,
\end{aligned}
\tag{7.116}
$$

where $\mu = m_{end}/((M+m)L)$ is a new dimensionless parameter that relates the mass that is added at the tip, m_{end}, to the total mass of the pipe and the fluid inside it. In this equation, $\delta(\xi - 1)$ is the Dirac delta function, which is zero everywhere expect at $\xi = 1$. All the terms due to the end mass have a delta function in them, because the end mass is located at $\xi = 1$ only. There are three new terms in the equation of motion due to the end mass: $\mu\delta(\xi - 1)\ddot{\eta}$ is an added inertia due to the end mass, and it is a linear term, implying that adding the end mass will influence the critical point for Hopf bifurcation as well, since the linear inertia terms are now different. The term in the fourth line of Eq. (7.116) is a new tension term, since adding a mass to the tip applies some tension to the beam, and the term in the last line of Eq. (7.116) is a nonlinear damping and inertia term. Note that the extra tension term also has nonlinear damping and inertia terms in it. If one adds an end mass to a hanging pipe, then gravity will also act, and it is easy to imagine that extra tension will act on the structure due to the gravity. In a more general version of the equation of motion, there are also several tension terms (both linear and nonlinear) due to the gravity; however, in Eq. (7.116), we have ignored gravity.

By considering only the linear part of Eq. (7.116), and solving the resulting eigenvalue problem, we can observe the influence of the added linear inertia term due to the added mass on the onset of oscillations. When an end mass is attached to the pipe, the critical value of the Hopf bifurcation decreases and the pipe starts oscillating at a lower flow velocity. As the value of the end mass is increased, the critical flow velocity decreases. This is shown in Fig. 7.15 for a pipe with $\beta = 0.2$. As the end mass parameter, μ, increases from 0 (no end mass) to $\mu = 0.2$, the critical value for Hopf bifurcation decreases from $u_{\mathrm{HB}} = 5.6$ to 3.4. This is a very dramatic decrease in the onset of instability by adding an end mass.

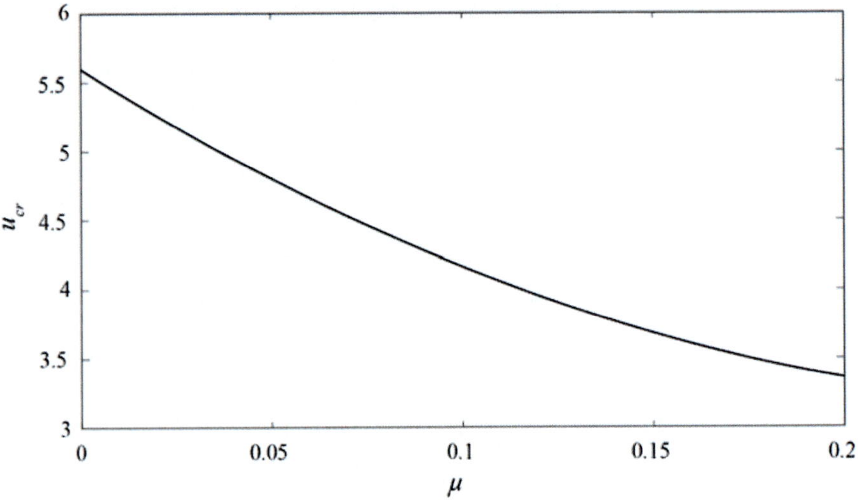

Fig. 7.15 The critical point for Hopf bifurcation versus the dimensionless end mass parameter, calculated for a cantilevered pipe with $\beta = 0.2$ and using 4 modes in the Galerkin discretization

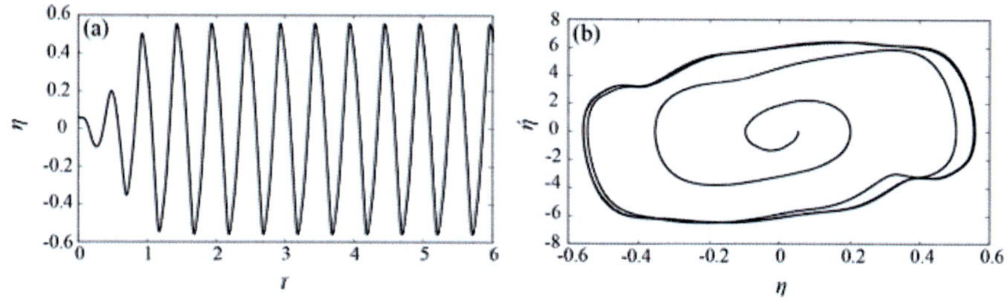

Fig. 7.16 (**a**) The time history and (**b**) the phase plane plot of the limit cycle response of a pipe conveying fluid with $\beta = 0.2$ and with an end mass parameter of $\mu = 0.05$

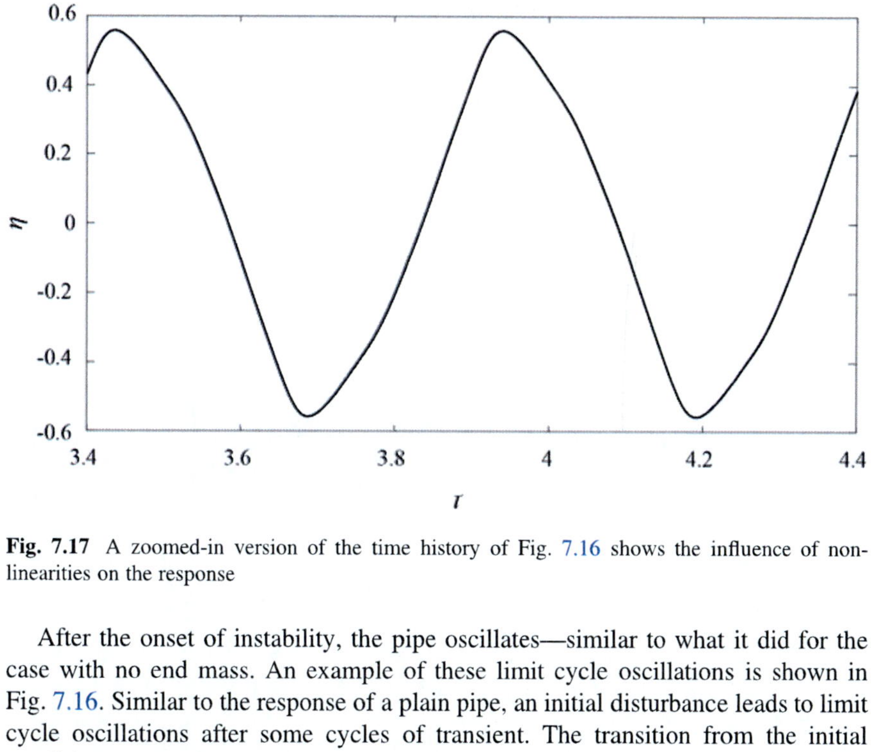

Fig. 7.17 A zoomed-in version of the time history of Fig. 7.16 shows the influence of nonlinearities on the response

After the onset of instability, the pipe oscillates—similar to what it did for the case with no end mass. An example of these limit cycle oscillations is shown in Fig. 7.16. Similar to the response of a plain pipe, an initial disturbance leads to limit cycle oscillations after some cycles of transient. The transition from the initial condition to the limit cycle response is clear in the phase plane of Fig. 7.16b. After spiraling in the clockwise direction for a few cycles, the response approaches the limit cycle, and stays on the limit cycle, which shows up as a closed curve in the phase plane plot. This closed curve is far from a circle or an ellipse, which is observed in the phase plane plot of a linear system. This deviation from an ellipse is a sign of nonlinearity in the system. These nonlinearities are also evident in the deviation of the amplitude plots from a regular sinusoidal motion, as clearly observed in the zoomed-in version of the time history in Fig. 7.17.

13 Three-Dimensional Behavior of a Pipe with an End Mass and Observation of Chaos

The behavior of a pipe with an end mass after the limit cycle oscillations is a mainly three-dimensional behavior and needs a three-dimensional model. Here we do not discuss such a three-dimensional model, and only review an example of the type of response that is observed in a cantilevered pipe with an end mass as the flow velocity is increased. A bifurcation diagram for a pipe conveying fluid with an end mass is shown in Fig. 7.18, where displacements in two perpendicular directions are plotted. After the first point of instability, the pipe undergoes planar flutter for $u = 5.4$ to 8.0 (region 1 in the figure). A phase plane of a periodic motion is a closed curve as shown in Fig. 7.19a. These planar oscillations become quasiperiodic oscillations, for $u = 8.0$–8.5 (region 2). The phase plane for quasiperiodic oscillations fills up a part of the plot and resembles a donut shape (Fig. 7.19b). A quasiperiodic response consists of two independent fundamental frequencies: a high frequency (one can think of this as a frequency of moving around a smaller diameter of a donut) and a low frequency (the frequency of moving around the larger diameter of a donut).

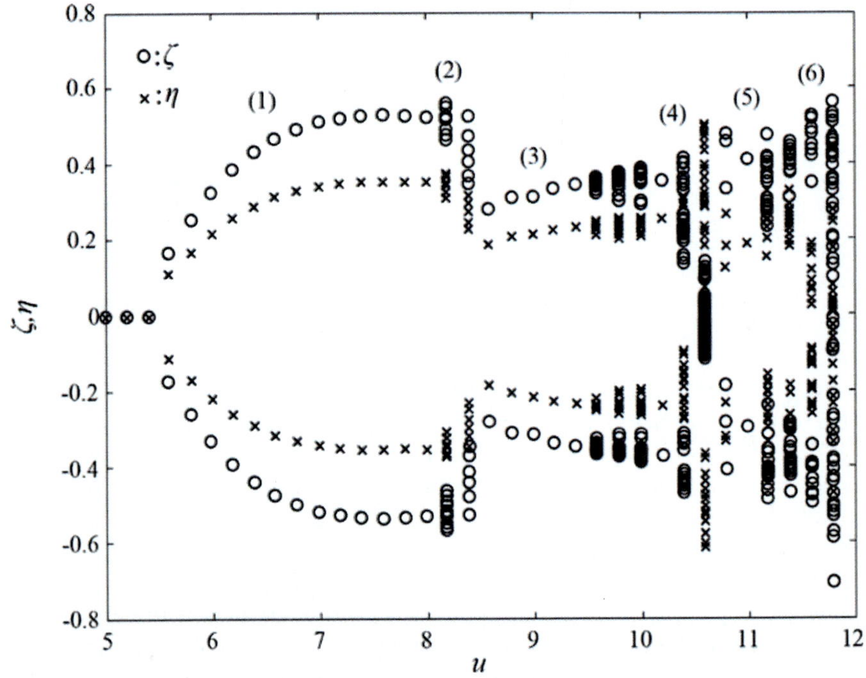

Fig. 7.18 Bifurcation diagram of the dimensionless amplitude of response in two perpendicular directions for a pipe with an end mass for $\beta = 0.145$ and $\mu = 0.1$. Note that gravity effects are also included. The numbers in parentheses refer to different types of response that have been observed. (From Modarres-Sadeghi et al., 2007)

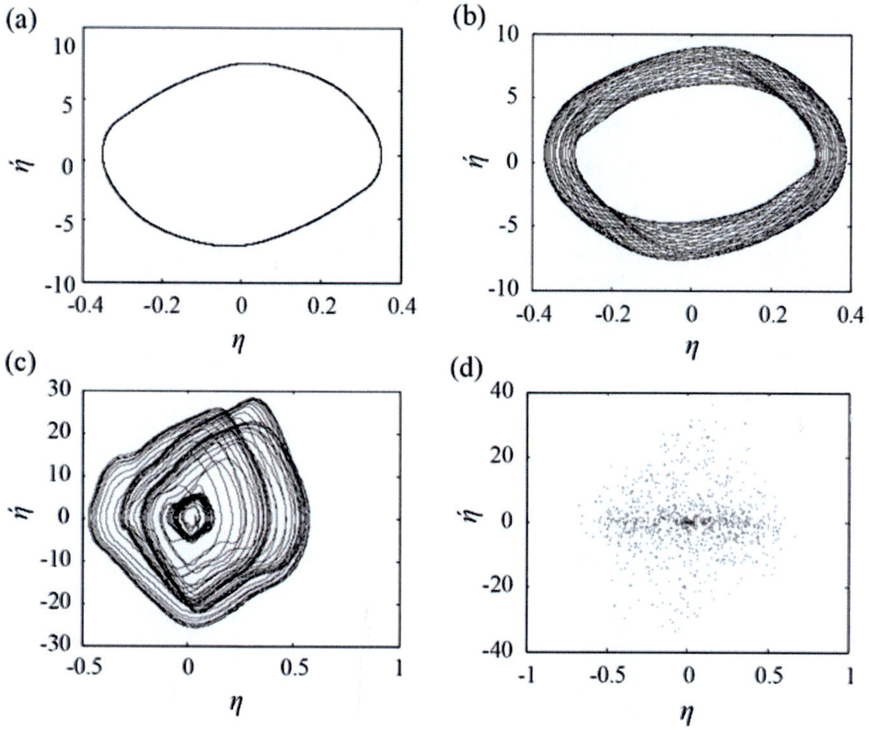

Fig. 7.19 Phase planes of the response at (**a**) $u = 7.8$ (periodic motion), (**b**) 8.2 (quasiperiodic motion), (**c**) 10.8 (quasiperiodic motion right before chaos), and (**d**) a Poincaré map at $u = 11.8$, where the response is chaotic. (From Modarres-Sadeghi et al., 2007)

Quasiperiodic oscillations are sometimes observed in a system before the system exhibits chaotic response as an independent variable is varied. For higher flow velocities of the present example, there is a sudden drop in the amplitude, which corresponds to higher-frequency periodic oscillations, where higher modes of the structure are excited for $u = 8.6–9.5$. At even higher flow velocities, three-dimensional oscillations are observed again, including regions of chaotic response. The phase plane shown in Fig. 7.19c is for a quasiperiodic motion at a flow velocity right before the motion becomes chaotic. The phase plane shows a degree of repeatability as some darker areas are observed where the system has passed more often than the other areas. As the flow velocity is increased slightly, this clear trend of the system becomes less clear, and trajectories in the phase plane fill up a part of the plot completely and all we will see will be a black region. To show this behavior more clearly, often a slice of the phase plane is shown for chaotic response of a system. This slice of the phase plane is called a Poincaré map. This map is shown for a sample chaotic response of the system of this example in Fig. 7.19d. The Poincaré map of a chaotic response looks like a cloud of points that tend to fill up a part of the plane—but not the entire plane. Poincaré maps are also used for periodic and

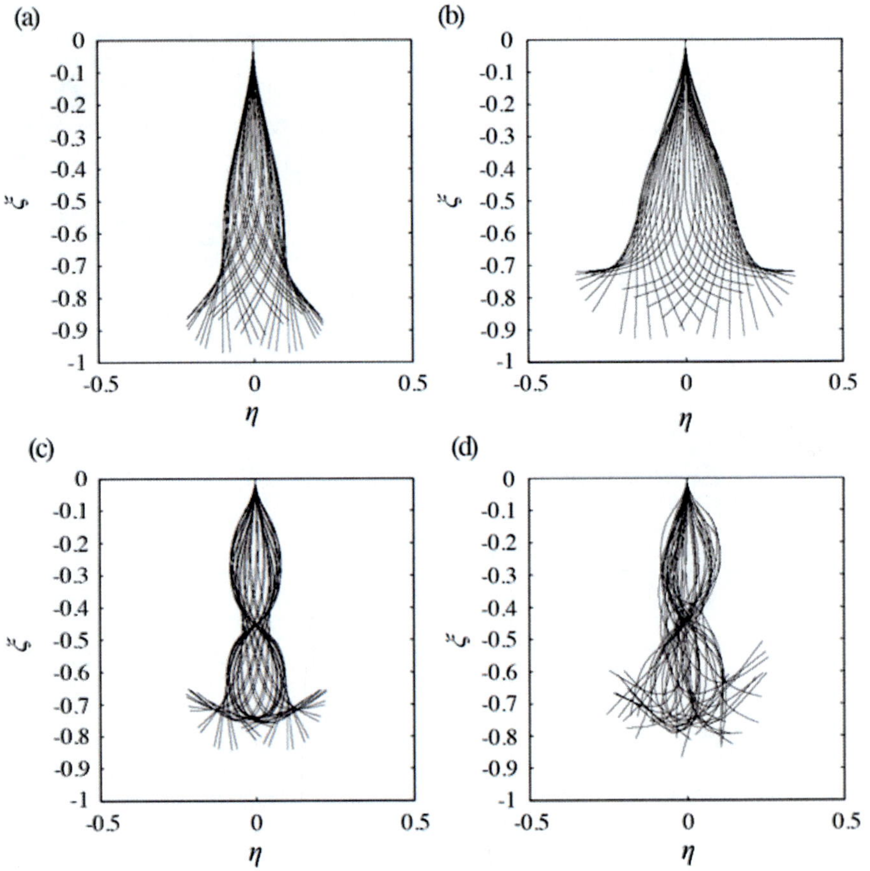

Fig. 7.20 Pipe behavior along its length for (**a**) $u = 6.0$ (**b**) $u = 7.8$ (**c**) $u = 9.0$ and (**d**) $u = 10$. (From Modarres-Sadeghi et al., 2007)

quasiperiodic motions in which cases they consist of a single point (or very limited number of points) for a periodic response, and a closed curve for a quasiperiodic motion.

Sample snapshots of the pipe's behavior along its length are given in Fig. 7.20. Initially, the pipe oscillates in its second mode (Fig. 7.20a). The amplitude of oscillations increases for increasing u (Fig. 7.20b), and then a mainly third mode response is observed (Fig. 7.20c), which results in a smaller amplitude of tip oscillations. At higher flow velocities, the pipe undergoes a torus instability, which is the instability that leads to a quasiperiodic response, and after going through a range of flow velocities for which the pipe exhibits quasiperiodic motion, chaos is observed (Fig. 7.20d).

This sequence of responses is an example of what we observe for a pipe with an end mass. Changing the system parameters changes the details of the response, but observation of chaotic response at higher flow velocities is common for different

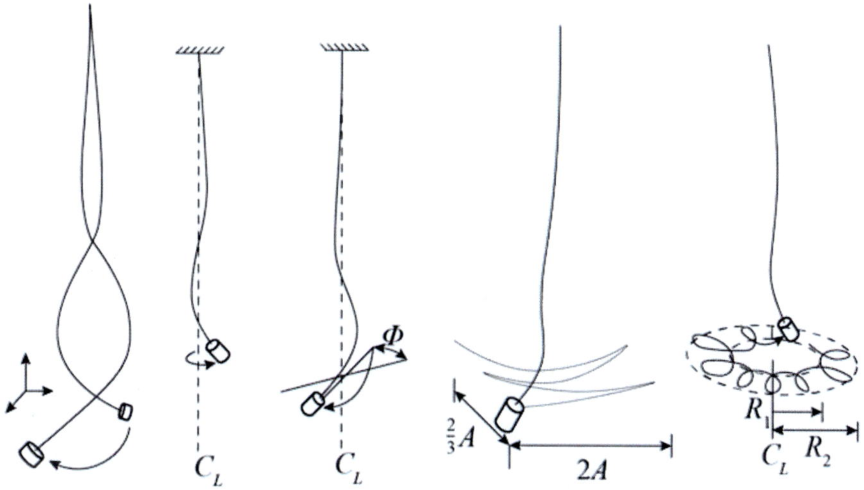

Fig. 7.21 Various oscillatory responses of a long pipe with an end mass studied by Copeland and Moon for various end masses and dimensionless flow velocities. (Reproduced from Copeland & Moon, 1992)

system parameters of a pipe with an end mass. A well-known example of behavior observed in a pipe with an end mass is a series of experiments that Copeland and Moon (1992) conducted. They used a very long pipe and attached relatively large end masses to its tip. For all the end masses that they studied, they observed that the pipe went through a series of three-dimensional oscillations and exhibited chaos at higher flow velocities. Before the chaotic response, they observed a series of three-dimensional patterns in the pipe, and they summarized them in a figure with hand-drawn sketches of the three-dimensional patterns (Fig. 7.21). The range of flow velocities from a static response to a chaotic response becomes wider for larger end masses, and more complicated three-dimensional patterns are observed. Copeland and Moon defined a dimensionless flow velocity, $u_g = U/(gL)^{1/2}$, in which the influence of gravity was also included, since gravity also becomes important when a pipe is hanging vertically. To reproduce Copeland and Moon's results numerically, we need a three-dimensional model that also includes the gravity effects. These models, and also numerical reproduction of these experimental results exist (Modarres-Sadeghi & Païdoussis, 2013).

Chapter 8
A Flexible Cylinder in Axial Flow

1 Introduction

In this chapter, we discuss another case of axial flow problems, a case of an external axial flow, in which a flexible beam with a circular cross section (a flexible cylinder) is placed in flow such that the flow is parallel to the long axis of the structure (Fig. 8.1). This is very similar to the pipe problem, but with an important difference: the flow is outside the structure, and as a result the amount of fluid that is affected by the motion of the structure is not known. In the pipe problem, since the flow was inside the structure, we knew exactly how much of the fluid is affected by the motion of the structure. In the external flow case, this amount is not known a priori. There will be an effective added mass of the fluid that the structure feels as it moves, and as a result there will be an added mass force acting on the structure. There will also be viscous forces that act from the flow on the structure, which unlike the pipe problem will not be canceled in the derivation of the equation of motion, and as a result the response will depend on the viscous forces as well. As we will see, the boundary conditions will have a significant influence on the response of the cylinder. In the case of a cantilevered cylinder, the forces that act on the tip become significant and influence the response of the system. Cylinders in axial flow are sometimes observed when a cylindrical structure is being towed in the ocean. Sometimes a structure that is originally perpendicular to the direction of flow bends, and a part of it becomes parallel to the flow. One of the very first studies on a flexible beam in water was the study conducted by Lighthill (1960) on the swimming of an eel. It was in this study that Lighthill proposed a method to derive the inviscid flow forces that act on a body due to the added mass of the flow, which we will use here in deriving the inviscid forces that act on a slender flexible beam in axial flow. In general, a cylinder in external axial flow is a canonical FSI problem, since it introduces ideas of how we can model external nonconservative flow forces that act on a flexible structure. Similar ideas can be used to model several other problems, including the problem of a flag flapping in the wind.

Y. Modarres-Sadeghi, *Introduction to Fluid-Structure Interactions*,
https://doi.org/10.1007/978-3-030-85884-1_8

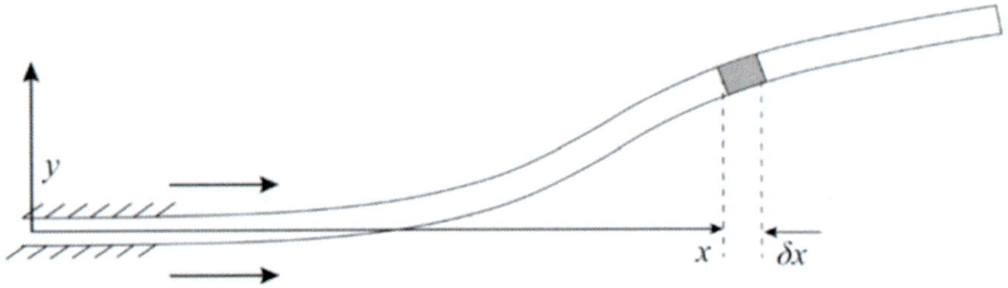

Fig. 8.1 Schematic of a flexible cylinder placed in axial flow

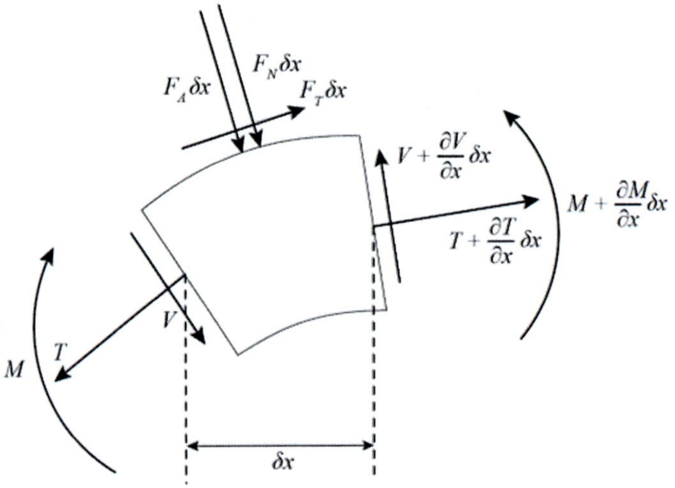

Fig. 8.2 Free-body diagram of an arbitrary element along the length of a flexible cylinder in axial flow

2 Derivation of the Equation of Motion

To derive the equation of motion for a cylinder placed in axial flow, we consider the free-body diagram of an arbitrary element of the cylinder (Fig. 8.2) and write Newton's second law in the x and y directions. First, we assume that the main displacement of the cylinder is in the direction perpendicular to its long axis, y, and the displacement in the x-direction is negligible compared with the displacement in the y-direction—they are an order of magnitude different from each other. In this derivation, we will follow the original work by Païdoussis (1973).

In the free-body diagram, there is a force, F_A, which acts on the structure in a direction perpendicular to its local curvature. This is the force that acts on the structure because the structure has to accelerate the fluid that is around it as the structure moves. This is then an added mass force—the same concept that we discussed in Chap. 3 for a cylinder in otherwise still fluid, and in Chap. 4, for the

VIV response of a flexibly mounted cylinder. To find this force, following Lighthill's original work, we have to find the acceleration of the element of the beam in the free-body diagram, and multiply that by the mass of the fluid that is moved because of the motion of that element. We assume this mass of the fluid to be $C_M M = C_M \rho A$, where C_M is an added mass coefficient, ρ is the density of fluid, and A is the cylinder's cross-sectional area.

To find the acceleration, we have to find the second derivative of the position, $y(x(t), t)$, with respect to time. Since y is a function of x, which itself is a function of t, we will have to find the second material derivative of y, as we discussed in Chap. 7. Then to find the added mass force, we multiply the second material derivate of y by the mass of the fluid that is displaced. This is a force that acts on the flow as a result of the motion of the structure, and according to Newton's third law, its reaction acts on the structure. Therefore, the inviscid force becomes:

$$\begin{aligned} F_A = C_M M \frac{D^2 y}{Dt^2} &= C_M M \left(\frac{\partial}{\partial t} + U \frac{\partial}{\partial x} \right)^2 y \\ &= C_M M \left(\frac{\partial^2 y}{\partial t^2} + 2U \frac{\partial^2 y}{\partial x \partial t} + U^2 \frac{\partial^2 y}{\partial x^2} \right). \end{aligned} \tag{8.1}$$

We model the viscous forces as two components, one normal to the structure, F_N, and one tangent to the structure, F_T:

$$F_N = \frac{1}{2} \rho D U C_N \frac{Dy}{Dt} = \frac{1}{2} \rho D U C_N \left(\frac{\partial y}{\partial t} + U \frac{\partial y}{\partial x} \right), \tag{8.2}$$

and

$$F_T = \frac{1}{2} \rho D U^2 C_T, \tag{8.3}$$

where D is the cylinder's diameter. Both these forces are written in a form very similar to the form that we normally use to write lift and drag forces, starting with a $\frac{1}{2} \rho D U^2$ and following by a coefficient that represents the flow force. In this case, C_N is the coefficient of the normal viscous force and C_T is the coefficient of the tangential viscous force. It is very clear from this description that these two coefficients have a significant influence on the viscous forces: by changing these coefficients we can change the viscous flow forces that act on the structure, the same way that we could change the lift and drag forces that act on a structure by changing the lift and drag coefficients.

The normal component of the viscous force is proportional to the material derivative of displacement, and therefore represents the flow-induced damping of the system. Besides the flow-induced damping, this normal component of the viscous force produces a term proportional to the slope of the structure: $\partial y / \partial x$.

The tangent component of the viscous force, F_T, acts as a friction force that is exerted on the structure.

Now we are ready to use the free-body diagram of Fig. 8.2 and derive the equation of motion. First, in the x-direction, we have:

$$\sum F_x = 0, \tag{8.4}$$

which gives us

$$(F_A\delta x + F_N\delta x)\frac{\partial y}{\partial x} + F_T\delta x - T + T + \frac{\partial T}{\partial x}\delta x = 0, \tag{8.5}$$

or

$$(F_A + F_N)\frac{\partial y}{\partial x} + F_T + \frac{\partial T}{\partial x} = 0. \tag{8.6}$$

Note that in Eq. (8.6), F_A and F_N are of the same order as y (in their definitions in Eqs. (8.2) and (8.3), we have the first and second derivatives of y with respect to time and space). When these forces are multiplied by $\partial y/\partial x$, then they produce terms of order y^2 (second-order terms), and since we are deriving a linear equation, the second-order term should be neglected. This then makes Eq. (8.6) take the form of

$$F_T + \frac{\partial T}{\partial x} = 0, \tag{8.7}$$

which is all we get in the x-direction.

Now, in the y-direction, where we have the main response of the cylinder, we have:

$$\sum F_y = ma_y,$$
$$\Rightarrow -V + \left(V + \frac{\partial V}{\partial x}\delta x\right) - F_A\delta x - F_N\delta x + F_T\frac{\partial y}{\partial x} - T\frac{\partial y}{\partial x}\bigg|_{x_1} + T\frac{\partial y}{\partial x}\bigg|_{x_2} = m\delta x\frac{\partial^2 y}{\partial t^2}. \tag{8.8}$$

Similar to what we did in deriving the equation of motion for the pipe, we relate $T\frac{\partial y}{\partial x}\big|_{x_2}$ to $T\frac{\partial y}{\partial x}\big|_{x_1}$ using Taylor expansion:

$$T\frac{\partial y}{\partial x}\bigg|_{x_2} = T\frac{\partial y}{\partial x}\bigg|_{x_1} + \frac{\partial}{\partial x}\left(T\frac{\partial y}{\partial x}\right)\delta x, \tag{8.9}$$

and plug it into Eq. (8.8) to get:

$$\frac{\partial V}{\partial x}\delta x - F_A\delta x - F_N\delta x + F_T\delta x\frac{\partial y}{\partial x} - T\frac{\partial y}{\partial x}\bigg|_{x_1} + T\frac{\partial y}{\partial x}\bigg|_{x_1} + \frac{\partial}{\partial x}\left(T\frac{\partial y}{\partial x}\right)\delta x$$
$$= m\delta x\frac{\partial^2 y}{\partial t^2}, \tag{8.10}$$

which simplifies to:

$$\frac{\partial V}{\partial x} - F_A - F_N + F_T\frac{\partial y}{\partial x} + \frac{\partial}{\partial x}\left(T\frac{\partial y}{\partial x}\right) = m\frac{\partial^2 y}{\partial t^2}. \tag{8.11}$$

We just found from the equation in the x-direction (Eq. (8.7)) that

$$F_T + \frac{\partial T}{\partial x} = 0, \tag{8.12}$$

because the viscous force in the tangential direction adds some tension to the system. By integrating Eq. (8.12) from x to L, we find:

$$\int_x^L F_T dx + T|_x^L = 0, \tag{8.13}$$

or

$$\int_x^L F_T dx + T_L - T = 0, \tag{8.14}$$

from which we can find T and plug it into Eq. (8.11) to get:

$$\frac{\partial V}{\partial x} - F_A - F_N + F_T\frac{\partial y}{\partial x} + \frac{\partial}{\partial x}\left(\left(\int_x^L F_T dx + T_L\right)\frac{\partial y}{\partial x}\right) = m\frac{\partial^2 y}{\partial t^2}, \tag{8.15}$$

and we expand the last term on the left-hand side to get:

$$\frac{\partial V}{\partial x} - F_A - F_N + F_T\frac{\partial y}{\partial x} + \frac{\partial}{\partial x}\left(\int_x^L F_T dx + T_L\right)\frac{\partial y}{\partial x} + \left(\int_x^L F_T dx + T_L\right)\frac{\partial^2 y}{\partial x^2}$$
$$= m\frac{\partial^2 y}{\partial t^2}, \tag{8.16}$$

which gives us two identical terms of $F_T\frac{\partial y}{\partial x}$ with opposite signs:

$$\frac{\partial V}{\partial x} - F_A - F_N + F_T\frac{\partial y}{\partial x} - F_T\frac{\partial y}{\partial x} + \left(\int_x^L F_T dx + T_L\right)\frac{\partial^2 y}{\partial x^2} = m\frac{\partial^2 y}{\partial t^2}, \quad (8.17)$$

and finally, after relating V to y as: $V = -EI\frac{\partial^3 y}{\partial x^3}$, and taking all terms to the same side, we find:

$$EI\frac{\partial^4 y}{\partial x^4} + F_A + F_N - \left(\int_x^L F_T dx + T_L\right)\frac{\partial^2 y}{\partial x^2} + m\frac{\partial^2 y}{\partial t^2} = 0. \quad (8.18)$$

We remember the expressions for F_A, F_N, and F_T in terms of y that we found earlier:

$$F_A = C_M M\left(\frac{\partial^2 y}{\partial t^2} + 2U\frac{\partial^2 y}{\partial x \partial t} + U^2\frac{\partial^2 y}{\partial x^2}\right), \quad (8.19)$$

$$F_N = \frac{1}{2}\rho D U C_N\left(\frac{\partial y}{\partial t} + U\frac{\partial y}{\partial x}\right), \quad (8.20)$$

and

$$F_T = \frac{1}{2}\rho D U^2 C_T, \quad (8.21)$$

and we plug them into Eq. (8.18) to get:

$$EI\frac{\partial^4 y}{\partial x^4} + C_M M\frac{\partial^2 y}{\partial t^2} + 2C_M MU\frac{\partial^2 y}{\partial x \partial t} + C_M MU^2\frac{\partial^2 y}{\partial x^2} + \frac{1}{2}\rho D U C_N\left(\frac{\partial y}{\partial t} + U\frac{\partial y}{\partial x}\right)$$
$$-\left[\frac{1}{2}\rho D U^2 C_T(L-x) + T_L\right]\frac{\partial^2 y}{\partial x^2} + m\frac{\partial^2 y}{\partial t^2} = 0, \quad (8.22)$$

or

$$EI\frac{\partial^4 y}{\partial x^4} + 2C_M MU\frac{\partial^2 y}{\partial x \partial t} + \frac{1}{2}\rho D U C_N\left(\frac{\partial y}{\partial t} + U\frac{\partial y}{\partial x}\right)$$
$$-\left[-C_M MU^2 + \frac{1}{2}\rho D U^2 C_T(L-x) + T_L\right]\frac{\partial^2 y}{\partial x^2} + (m + C_M M)\frac{\partial^2 y}{\partial t^2} = 0. \quad (8.23)$$

This is almost the final form of the equation of motion. We only need to decide on T_L, which we will do soon. Before that, we consider each term and we discuss their physical significance. The first term, $EI\partial^4 y/\partial x^4$, is the force due to the flexural rigidity of the structure, the structure's stiffness. The second term, $2C_M MU\partial^2 y/\partial x\partial t$, is the Coriolis term, similar to the Coriolis term that we observed in the pipe equation. The difference is that the added mass has a coefficient, C_M, which could be different from 1. The term $C_M MU^2\partial^2 y/\partial x^2$ is the centripetal term, again similar to the term observed in the pipe equation, but with an extra C_M in it, and the last term, $(m + C_M M)\partial^2 y/\partial t^2$, is the inertia term, including the added mass effect. There are also some terms that we did not observe in the pipe equation, but they appear in the cylinder's equation, and those are the terms due to the viscous forces—terms with C_N and C_T in them. The term $\frac{1}{2}\rho DUC_N(\partial y/\partial t + U\partial y/\partial x)$ summarizes the influence of the normal component of the viscous force, F_N, in the final equation of motion: a flow-induced damping term, and a term proportional to the slope of the cylinder. The tangential component of the viscous force appears in the tension term, $-\frac{1}{2}\rho DU^2 C_T(L - x)\partial^2 y/\partial x^2$, and adds to the tension that the cylinder feels at its tip, T_L. This tension at the tip will be different for a cylinder supported at both ends and a cantilevered cylinder. If the cylinder is cantilevered, the tension at the tip is written in the form of a base drag:

$$T_L = \frac{1}{2}\rho D^2 U^2 C_b, \tag{8.24}$$

in which C_b is a coefficient that determines the magnitude of this force. For a cylinder supported at both ends, however, we can write the tension at the end as:

$$T_L = -\frac{1}{2}\rho DU^2 C_T \frac{L}{2} + \overline{T}, \tag{8.25}$$

which consists of a term proportional to C_T that acts at the end of the cylinder as a reaction force to the tangential component of the viscous force, and an externally applied tension, $\overline{T}$.

By considering these two versions of T_L, we can find that for a cantilevered cylinder, the equation of motion becomes:

$$\begin{aligned} &EI\frac{\partial^4 y}{\partial x^4} + 2C_M MU\frac{\partial^2 y}{\partial x\partial t} + \frac{1}{2}\rho DUC_N\left(\frac{\partial y}{\partial t} + U\frac{\partial y}{\partial x}\right) \\ &-\left[-C_M MU^2 + \frac{1}{2}\rho DU^2 C_T(L - x) + \frac{1}{2}\rho D^2 U^2 C_b\right]\frac{\partial^2 y}{\partial x^2} + (m + C_M M)\frac{\partial^2 y}{\partial t^2} = 0, \end{aligned} \tag{8.26}$$

and for a beam supported at both ends, it becomes:

$$EI\frac{\partial^4 y}{\partial x^4}+2C_M MU\frac{\partial^2 y}{\partial x\partial t}+\frac{1}{2}\rho DUC_N\left(\frac{\partial y}{\partial t}+U\frac{\partial y}{\partial x}\right)$$
$$-\left[-C_M MU^2+\frac{1}{2}\rho DU^2 C_T\left(\frac{L}{2}-x\right)+\overline{T}\right]\frac{\partial^2 y}{\partial x^2}+(m+C_M M)\frac{\partial^2 y}{\partial t^2}=0. \tag{8.27}$$

We can write the two equations together, by introducing a parameter δ, which is equal to 1 for a beam supported at both ends and zero for a beam with a free end, and get the combined version of the equation of motion for a cylinder in axial flow as:

$$EI\frac{\partial^4 y}{\partial x^4}+2C_M MU\frac{\partial^2 y}{\partial x\partial t}+\frac{1}{2}\rho DUC_N\left(\frac{\partial y}{\partial t}+U\frac{\partial y}{\partial x}\right)$$
$$-\left[-C_M MU^2+\frac{1}{2}\rho DU^2 C_T(L-x)+\left(-\frac{1}{2}\rho DU^2 C_T\frac{L}{2}+\overline{T}\right)\delta-(\delta-1)\frac{1}{2}\rho D^2U^2C_b\right]\frac{\partial^2 y}{\partial x^2}$$
$$+(m+C_M M)\frac{\partial^2 y}{\partial t^2}=0. \tag{8.28}$$

Note that in this equation, we have assumed that the cylinder is horizontal, and gravity does not act in the direction of its motion. If the gravity effects are also included, then some extra tension terms will show up in the equation of motion.

3 Dimensionless Equation of Motion

The first step before solving the equation of motion for a linear cylinder is to make the equation dimensionless. To do so, similar to what we did for a pipe conveying fluid in Chap. 7, we introduce dimensionless displacements and time as:

$$\eta=\frac{y}{L}, \tag{8.29}$$

$$\xi=\frac{x}{L}, \tag{8.30}$$

and

$$\tau=\left(\frac{EI}{m+M}\right)^{½}\frac{1}{L^2}t, \tag{8.31}$$

and we plug these parameters into the equation of motion (Eq. (8.28)) to get:

$$\frac{\partial^4(\eta L)}{\partial(\xi L)^4}+2C_M\frac{MU}{EI}\frac{\partial^2(\eta L)}{\partial(\xi L)\partial\left(\left(\frac{m+M}{EI}\right)^{\frac{1}{2}}L^2\tau\right)}+\frac{1}{2}\rho DUC_N\frac{1}{EI}\left(\frac{\partial(\eta L)}{\partial\left(\left(\frac{m+M}{EI}\right)^{\frac{1}{2}}L^2\tau\right)}+U\frac{\partial(\eta L)}{\partial(\xi L)}\right)$$
$$-\left[-C_MMU^2+\frac{1}{2}\rho DU^2C_T(L-x)+\left(-\frac{1}{2}\rho DU^2C_T\frac{L}{2}+\overline{T}\right)\delta-(\delta-1)\frac{1}{2}\rho D^2U^2C_b\right]\frac{1}{EI}\frac{\partial^2(\eta L)}{\partial(\xi L)^2}$$
$$+(m+C_MM)\frac{1}{EI}\frac{\partial^2(\eta L)}{\partial\left(\left(\frac{m+M}{EI}\right)^{\frac{1}{2}}L^2\tau\right)^2}=0, \tag{8.32}$$

and after some simplifications and multiplying all terms by L^3:

$$\frac{\partial^4\eta}{\partial\xi^4}+2LC_M\frac{MU}{EI}\left(\frac{EI}{m+M}\right)^{\frac{1}{2}}\frac{\partial^2\eta}{\partial\xi\partial\tau}+\frac{1}{2}\rho DUC_N\left(\frac{EI}{m+M}\right)^{\frac{1}{2}}\frac{L^2}{EI}\frac{\partial\eta}{\partial\tau}+\frac{1}{2}\rho DU^2C_N\frac{L^3}{EI}\frac{\partial\eta}{\partial\xi}$$
$$-L^2\left[-C_MMU^2+\frac{1}{2}\rho DU^2C_T(L-x)+\left(-\frac{1}{2}\rho DU^2C_T\frac{L}{2}+\overline{T}\right)\delta-(\delta-1)\frac{1}{2}\rho D^2U^2C_b\right]\frac{1}{EI}\frac{\partial^2\eta}{\partial\xi}$$
$$+\left(1+\frac{C_M-1}{m+M}M\right)\frac{\partial^2\eta}{\partial\tau^2}=0. \tag{8.33}$$

By re-arranging the coefficients of different terms in this equation, we find:

$$\frac{\partial^4\eta}{\partial\xi^4}+2C_M\frac{M^{\frac{1}{2}}UL}{(EI)^{\frac{1}{2}}}\left(\frac{M}{m+M}\right)^{\frac{1}{2}}\frac{\partial^2\eta}{\partial\xi\partial\tau}+$$
$$+\frac{1}{2}\rho D\frac{\frac{\pi}{4}\rho D^2}{\frac{\pi}{4}\rho D^2}UC_N\frac{1}{M^{\frac{1}{2}}}\left(\frac{M}{m+M}\right)^{\frac{1}{2}}\frac{L^2}{(EI)^{\frac{1}{2}}}\frac{\partial\eta}{\partial\tau}+\frac{1}{2}\rho D\frac{\frac{\pi}{4}\rho D^2}{\frac{\pi}{4}\rho D^2}U^2C_N\frac{L^3}{EI}\frac{\partial\eta}{\partial\xi}$$
$$-\left[-C_M\frac{ML^2U^2}{EI}+\frac{1}{2}\rho D\frac{L^2U^2}{EI}C_T(L-x)+\left(-\frac{1}{2}\rho D\frac{L^2U^2}{EI}C_T\frac{L}{2}+\frac{\overline{T}L^2}{EI}\right)\delta-(\delta-1)\frac{1}{2}\rho D^2\frac{U^2L^2}{EI}C_b\right]\frac{\partial^2\eta}{\partial\xi^2}$$
$$+\left(1+(C_M-1)\frac{M}{m+M}\right)\frac{\partial^2\eta}{\partial\tau^2}=0, \tag{8.34}$$

which contains the dimensionless parameters that we had seen in the dimensionless pipe problem, $u=\sqrt{M/(EI)}UL$ and $\beta=M/(m+M)$, the dimensionless parameter we had observed in the Holmes equation, $\Pi_0=\overline{T}L^2/(EI)$, and some new dimensionless parameters, that we define as:

$$\varepsilon=\frac{L}{D},c_N=\frac{C_N}{\frac{\pi}{4}},c_T=\frac{C_T}{\frac{\pi}{4}},c_B=\frac{C_B}{\frac{\pi}{4}}, \tag{8.35}$$

and then we can write the dimensionless version of the equation as

$$\frac{\partial^4\eta}{\partial\xi^4} + 2C_M u\beta^{\frac{1}{2}}\frac{\partial^2\eta}{\partial\xi\partial\tau} + \frac{1}{2}uc_N\beta^{\frac{1}{2}}\varepsilon\frac{\partial\eta}{\partial\tau} + \frac{1}{2}u^2c_N\varepsilon\frac{\partial\eta}{\partial\xi}$$
$$- \left[-C_M u^2 + \frac{1}{2}u^2c_T\varepsilon(1-\xi) + \left(-\frac{1}{2}u^2c_T\frac{\varepsilon}{2} + \Pi_0\right)\delta - (\delta-1)\frac{1}{2}u^2c_b\right]\frac{\partial^2\eta}{\partial\xi^2}$$
$$+ (1 + (C_M - 1)\beta)\frac{\partial^2\eta}{\partial\tau^2} = 0. \tag{8.36}$$

Now that we have the dimensionless equation of motion, we are ready to solve it.

4 Method of Solution

To solve the linear equation of motion of a cylinder in axial flow (Eq. (8.36)), we use the Galerkin technique, again similar to what we did to solve the pipe equation in Chap. 7, by assuming the general solution to be the same as Eq. (7.60) and plugging it into the equation of motion to get:

$$\sum_{j=1}^{N}\phi_j''''\,q_j + 2C_M u\beta^{\frac{1}{2}}\sum_{j=1}^{N}\phi'_j\dot{q}_j + \frac{1}{2}uc_N\beta^{\frac{1}{2}}\,\varepsilon\sum_{j=1}^{N}\phi_j\dot{q}_j + \frac{1}{2}u^2c_N\varepsilon\sum_{j=1}^{N}\phi'_jq_j$$
$$-\left[-C_M u^2 + \frac{1}{2}u^2c_T\varepsilon(1-\xi) + \left(-\frac{1}{2}u^2c_T\frac{\varepsilon}{2} + \Pi_0\right)\delta - (\delta-1)\frac{1}{2}u^2c_b\right]\sum_{j=1}^{N}\phi''_jq_j$$
$$+(1 + (C_M - 1)\beta)\sum_{j=1}^{N}\phi_j\ddot{q}_j = 0. \tag{8.37}$$

After multiplying all terms by ϕ_i, integrating from 0 to 1, and collecting the stiffness, damping, and inertia terms, we find:

$$\sum_{j=1}^{N}\Bigg[\Bigg(\int_0^1\phi_i\phi_j''''\mathrm{d}\xi + \frac{1}{2}u^2c_N\varepsilon\int_0^1\phi_i\phi'_j\mathrm{d}\xi - \Big[-C_M u^2 + \frac{1}{2}u^2c_T\varepsilon(1-\xi) + \left(-\frac{1}{2}u^2c_T\frac{\varepsilon}{2} + \Pi_0\right)\delta$$
$$-(\delta-1)\frac{1}{2}u^2c_b\Big]\int_0^1\phi_i\phi''_j\mathrm{d}\xi\Bigg)q_j + \left(2C_M u\beta^{\frac{1}{2}}\int_0^1\phi_i\phi'_j\mathrm{d}\xi + \frac{1}{2}uc_N\beta^{\frac{1}{2}}\,\varepsilon\int_0^1\phi_i\phi_j\mathrm{d}\xi\right)\dot{q}_j$$
$$+(1 + (C_M - 1)\beta)\int_0^1\phi_i\phi_j\mathrm{d}\xi\ddot{q}_j\Bigg] = 0, \tag{8.38}$$

which can be written as an N-degree of freedom mass-spring-damper system, where N is the number of modes used in discretization:

$$K_{ij}q_j + C_{ij}\dot{q}_j + M_{ij}\ddot{q}_j = 0. \tag{8.39}$$

This is the same format for the set of ODEs that we found when we solved the linear equation of motion for a pipe in Chap. 7. From here, by picking the appropriate general solution for this set of ODEs, we will obtain an eigenvalue problem, which we can solve for the parameters of a cylinder in axial flow and for different values of u to investigate how the cylinder's behavior changes as the flow velocity is increased.

5 Linear Behavior of a Cylinder Supported at Both Ends

We start discussing the behavior of a flexible cylinder by focusing on a simply supported cylinder and by inspecting how the cylinder's stability changes as the flow velocity is increased. Figure 8.3 shows the real and imaginary parts of the eigenvalues for a cylinder in axial flow with $\beta = 0.1$, $\varepsilon = 10$, $\Pi_0 = 0$, $C_M = 1$, and $C_n = C_t = 0.1$, and since the cylinder is supported at both ends, $\delta = 1$. It is seen from the eigenvalue plots that the real part is zero initially, and it stays zero up to $u_b = 3.14 = \pi$. The real part of the eigenvalue is proportional to the damping of the structure, and it is zero initially, because we have assumed that there is no structural damping. The imaginary parts of the eigenvalues, which are proportional to the natural frequencies of the system, decrease with increasing flow velocity until $u_b = 3.14 = \pi$, where the frequency of the first mode becomes zero, indicating that the system has buckled in its first mode. This behavior is very similar to the behavior of a simply supported pipe conveying fluid. As we saw in Chap. 7, a simply supported pipe buckles at $u_b = 3.14 = \pi$. And here, we observe the same behavior for a simply supported cylinder in axial flow. Similar to what we discussed in the case of a pipe conveying fluid, the system's behavior for flow velocities larger than the critical cannot be reliably predicted by the linear model, because the cylinder could undergo large displacements after the first point of instability, which will make the assumptions of a linear model invalid.

The response of a clamped-clamped cylinder in flow is qualitatively very similar to a simply supported cylinder as well, as shown in Fig. 8.4. From these plots, it is clear that the first mode of the structure becomes unstable at $u_b = 6.28 = 2\pi$ and the cylinder buckles at its first mode. Interestingly, this is also the same critical value for buckling as the critical value for a pipe conveying fluid and clamped at its both ends as we discussed in Chap. 7. One of the parameters that influences the onset of instability of a flexible cylinder supported at its both ends is the added mass coefficient, C_M. In the results of Figs. 8.3 and 8.4 we assume $C_M = 1$. For a pipe conveying fluid, C_M is always equal to 1, because the added mass can be calculated

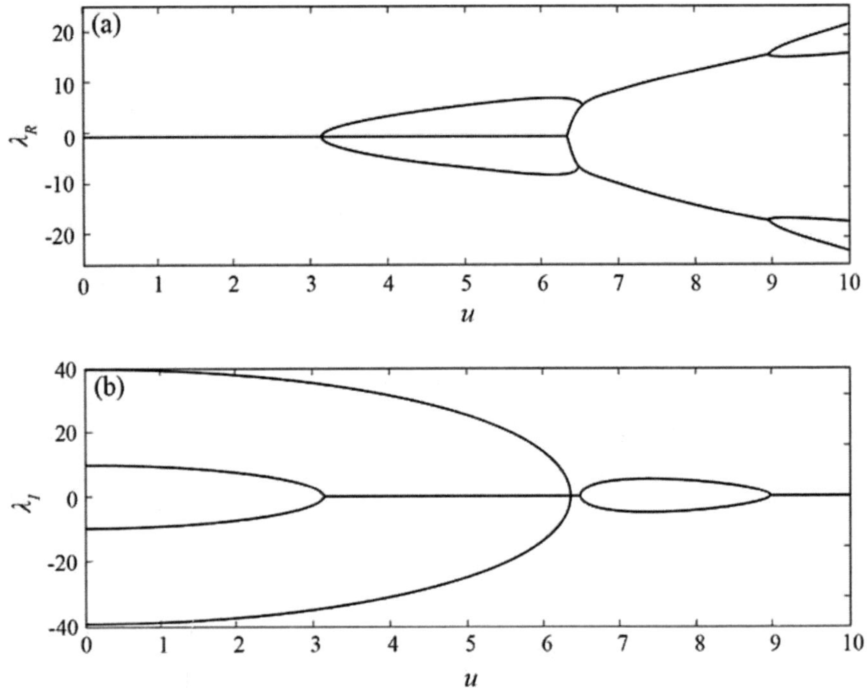

Fig. 8.3 (**a**) Real and (**b**) imaginary parts of the eigenvalues for a simply supported cylinder in axial flow for $\beta = 0.1$, $\varepsilon = 10$, $\Pi_0 = 0$, $C_M = 1$, and $C_n = C_t = 0.1$

precisely, as the area that the moving fluid covers is exactly equal to the inner size of the pipe. However, for a cylinder, the assumption that $C_M = 1$ is based on the potential flow calculations. We have found exactly the same critical values for the onset of buckling in a cylinder supported at both ends and a pipe supported at both ends, because we have used $C_M = 1$ for the cylinder as well. For an external flow case, however, the added mass coefficient could be different from 1. If we change the value of C_M, then the critical flow velocity for buckling changes as well. For example, for a $C_M = 1.2$, the critical flow velocity for buckling decreases to $u_b = 2.9$ for a simply supported cylinder. Although the critical value for the onset of buckling can change by changing the added mass coefficient, the type of instability will not change, and a cylinder supported at its both ends always buckles first, before it undergoes any other type of instability.

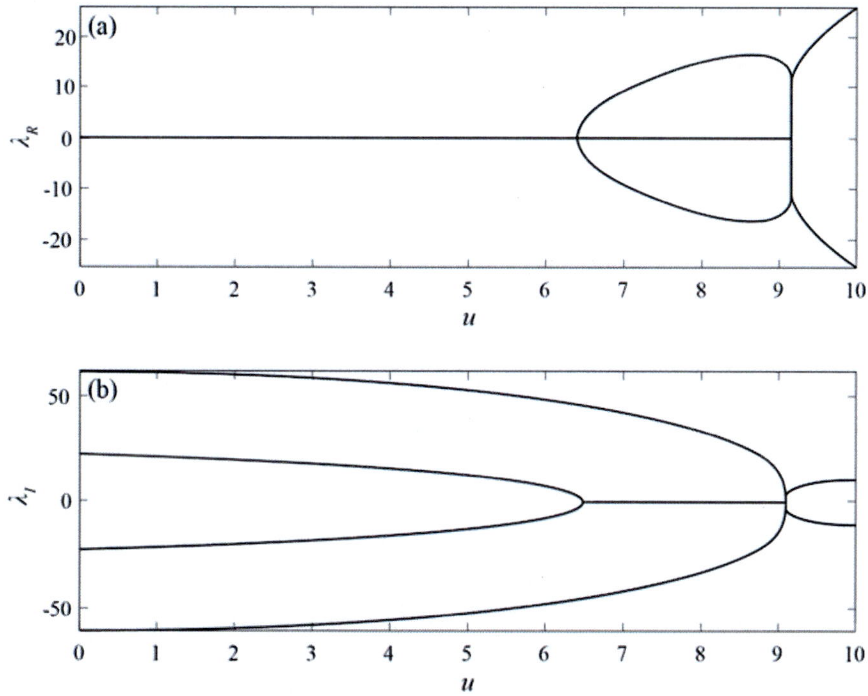

Fig. 8.4 (**a**) Real and (**b**) imaginary parts of the eigenvalues for a clamped-clamped cylinder in axial flow, with the same parameters as in Fig. 8.3 ($\beta = 0.1$, $\varepsilon = 10$, $\Pi_0 = 0$, $C_M = 1$, and $C_n = C_t = 0.1$)

6 A Cantilevered Cylinder

For a cantilevered cylinder, the geometry of the tip has a significant influence on the observed dynamics. If the end is not blunt, i.e., cut completely perpendicular to the long axis of the cylinder, then there will be flow forces that act on the tip in the direction of motion, and therefore the boundary condition for a cantilevered cylinder cannot be a shear force of zero, but a nonzero shear force will act at the tip. To find this force, following Païdoussis (1973), we consider a conical shape for the tip such that its diameter starts with the diameter of the cylinder itself and goes down to zero over a small length of l, where $l << L$. Then by considering the free-body diagram of a slice of this conical shape (Fig. 8.5), we can relate the shear force at the boundary of the cylinder to the flow forces that act on its conical end. From the free-body diagram, in the y-direction:

$$\begin{aligned} &\sum F_y = ma_y \\ &\Rightarrow -V + \left(V + \frac{\partial V}{\partial x}dx\right) - fF_A dx = mdx\frac{\partial^2 y}{\partial t^2}, \end{aligned} \tag{8.40}$$

where F_A is the added mass force that acts on the conical shape, and can be written as

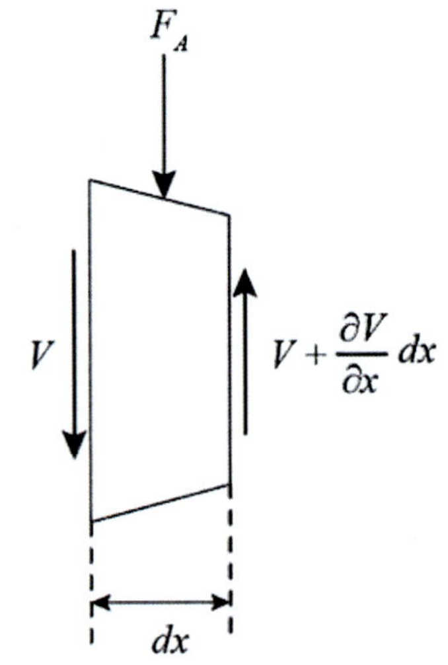

Fig. 8.5 Free-body diagram of an element of the tip of a cantilevered cylinder

$$F_A = \frac{D}{Dt}(C_M M v) = \left(\frac{\partial}{\partial t} + U\frac{\partial}{\partial x}\right)(C_M M v), \tag{8.41}$$

and f is a parameter smaller than 1 that reflects the deviation of the added mass force at the tip, due to its inclination, from the ideal two-dimensional added mass force, for which $f = 1$. Here, we assume that the cylinder's velocity does not change over the small length of the conical shape, and therefore v is only a function of time. Therefore,

$$F_A = C_M M\frac{\partial v}{\partial t} + C_M U v\frac{\partial M}{\partial x} = C_M M\frac{\partial^2 y}{\partial t^2} + C_M U v\frac{\partial M}{\partial x}. \tag{8.42}$$

If we plug this version of the inviscid force into Eq. (8.40), we get:

$$\frac{\partial V}{\partial x}dx - fC_M M\frac{\partial^2 y}{\partial t^2}dx - fC_M U v\frac{\partial M}{\partial x}dx = mdx\frac{\partial^2 y}{\partial t^2}. \tag{8.43}$$

Then we integrate this over the entire length of the conical shape to get:

$$\int_{L-l}^{L}\left(\frac{\partial V}{\partial x}dx - fC_M M\frac{\partial^2 y}{\partial t^2}dx - fC_M U v\frac{\partial M}{\partial x}dx\right) = \int_{L-l}^{L}\left(mdx\frac{\partial^2 y}{\partial t^2}\right), \tag{8.44}$$

or

$$V\Big|_{L-l}^{L} - fC_M\frac{\partial^2 y}{\partial t^2}\int_{L-l}^{L} Mdx - fC_M U v\int_{L-l}^{L}\frac{\partial M}{\partial x}dx = \frac{\partial^2 y}{\partial t^2}\int_{L-l}^{L}(mdx). \tag{8.45}$$

We plug $M = \rho A(x)$ and $m = \rho_s A(x)$, where ρ_s is the density of the cylinder, and $A(x)$ is how the cross-sectional area of the conical tip changes, to get:

$$V\Big|_{L-l}^{L} - fC_M \frac{\partial^2 y}{\partial t^2} \int_{L-l}^{L} \rho A(x)dx - fC_M Uv \int_{L-l}^{L} \frac{\partial M}{\partial x} dx = \frac{\partial^2 y}{\partial t^2} \int_{L-l}^{L} (\rho_s A(x)dx), \tag{8.46}$$

or

$$V\Big|_{L-l}^{L} - fC_M \frac{\partial^2 y}{\partial t^2} \rho \int_{L-l}^{L} A(x)dx - fC_M Uv M\Big|_{L-l}^{L} = \frac{\partial^2 y}{\partial t^2} \rho_s \int_{L-l}^{L} (A(x)dx). \tag{8.47}$$

We then set:

$$\int_{L-l}^{L} A(x)dx = Ax_E \tag{8.48}$$

and re-write the equation as:

$$V_L - V_{L-l} - fC_M \frac{\partial^2 y}{\partial t^2} \rho A x_E - fC_M Uv(M_L - M_{L-l}) = \frac{\partial^2 y}{\partial t^2} \rho_s A x_E. \tag{8.49}$$

Note that here $M_{L-l} \equiv M$ and $M_L \equiv 0$, and also $V_{L-l} = -EI \frac{\partial^3 y}{\partial x^3}$ and $V_L = 0$. Therefore:

$$EI \frac{\partial^3 y}{\partial x^3} - fC_M \frac{\partial^2 y}{\partial t^2} \rho A x_E + fC_M Uv M - \frac{\partial^2 y}{\partial t^2} \rho_s A x_E = 0, \tag{8.50}$$

or

$$EI \frac{\partial^3 y}{\partial x^3} = -fC_M UM\left(\frac{\partial y}{\partial t} + U\frac{\partial y}{\partial x}\right) + (fC_M M + m)x_E \frac{\partial^2 y}{\partial t^2}, \tag{8.51}$$

which is the shear force at the boundary of the beam.

First, we use the dimensionless variables that we introduced before and make this relation dimensionless:

$$\begin{aligned} EI \frac{1}{L^2} \frac{\partial^3 \eta}{\partial \xi^3} &= -fC_M UM\left(\frac{EI}{m+M}\right)^{½} \frac{1}{L} \frac{\partial \eta}{\partial \tau} - fC_M MU^2 \frac{\partial \eta}{\partial \xi} \\ &\quad + (fM + m)x_E\left(\frac{EI}{m+M}\right) \frac{1}{L^3} \frac{\partial^2 \eta}{\partial \tau^2}, \end{aligned} \tag{8.52}$$

or

$$\frac{\partial^3\eta}{\partial\xi^3} = -fC_M\left(\frac{M}{m+M}\right)^{1/2}\frac{M^{\frac{1}{2}}UL}{(EI)^{\frac{1}{2}}}\frac{\partial\eta}{\partial\tau} - fC_M\frac{MU^2L^2}{EI}\frac{\partial\eta}{\partial\xi} + \left(1+\frac{M}{m+M}(fC_M-1)\right)\frac{x_E}{L}\frac{\partial^2\eta}{\partial\tau^2}, \tag{8.53}$$

which can be written in terms of the dimensionless parameters of the system as:

$$\frac{\partial^3\eta}{\partial\xi^3} = -fC_M\beta^{1/2}u\frac{\partial\eta}{\partial\tau} - fC_Mu^2\frac{\partial\eta}{\partial\xi} + (1+\beta(fC_M-1))x_e\frac{\partial^2\eta}{\partial\tau^2}, \tag{8.54}$$

where

$$x_e = \frac{x_E}{L}. \tag{8.55}$$

This equation then can be added to the equation of motion using a Dirac delta function, which is zero everywhere, except at the tip, i.e., $\xi = 1$. The implication of using a Dirac delta function in these terms is that when we discretize them using the Galerkin technique, we use

$$\eta = \phi_j(1)q_j, \tag{8.56}$$

in which the mode shape is calculated only at the tip, and is only one number. By plugging Eq. (8.56) into Eq. (8.54), we get:

$$\frac{\partial^3\eta}{\partial\xi^3} = -fC_M\beta^{1/2}u\phi_j(1)\dot{q}_j - fC_Mu^2\phi'_j(1)q_j + (1+\beta(fC_M-1))x_e\phi_j(1)\ddot{q}_j. \tag{8.57}$$

After adding the right-hand side to the equation of motion, we will have to calculate the contribution of each mode in the total response. To do so, we multiply the equation by $\phi_i(1)$ to get:

$$-fC_M\beta^{1/2}u\phi_i(1)\phi_j(1)\dot{q}_j - fC_Mu^2\phi_i(1)\phi'_j(1)q_j + (1+\beta(fC_M-1))x_e\phi_i(1)\phi_j(1)\ddot{q}_j, \tag{8.58}$$

which gives us an additional damping term, an additional stiffness term, and an additional inertia term, respectively. These terms need to be added to the damping, stiffness, and inertia terms of the original equation of motion. The influence of the tip then is felt in all aspects of the response, since the flow forces that act at the tip influence stiffness, damping, and inertia of the system. In what follows, we will look at a typical linear response of a cantilevered cylinder in axial flow.

7 Linear Results for a Cantilevered Cylinder

A sample case of the linear results for a cylinder in axial flow is given in Fig. 8.6 for $\beta = 0.5$, $\varepsilon = 10$, $\Pi_0 = 0$, $C_n = C_t = 0.1$, $f = 0.8$, $C_b = 0$, $x_e = 0.01$, and since the end is free, $\delta = 0$. The results show that the cylinder buckles to its first mode at a critical flow velocity of $u = 1.9$. This is similar to the buckling of a cylinder supported at both ends, but at a different critical value. For a cantilevered cylinder, the onset of buckling depends on several system parameters.

One of the parameters that influences the onset of buckling is f. A lower value of f leads to a higher critical flow velocity for buckling. For example, a cylinder with the parameters of Fig. 8.6 buckles at $u = 2.4$ for $f = 0.5$, and $u = 2.6$ for $f = 0.3$. Parameter f reflects the deviation of the added mass force of the tip from the added mass force of the rest of the cylinder. The added mass changes as the diameter of the cylinder changes at its tip. A smaller f means a smaller added mass and a higher frequency of the system at any given flow velocity. Then, as the flow velocity is increased, it takes longer for the frequency to go to zero, and as a result the onset of buckling is delayed (Fig. 8.7).

Other parameters that influence the onset of buckling significantly are c_N and c_T. A cylinder with the parameters of Fig. 8.6 buckles at $u = 1.85$ for $c_N = c_T = 0.01$ and

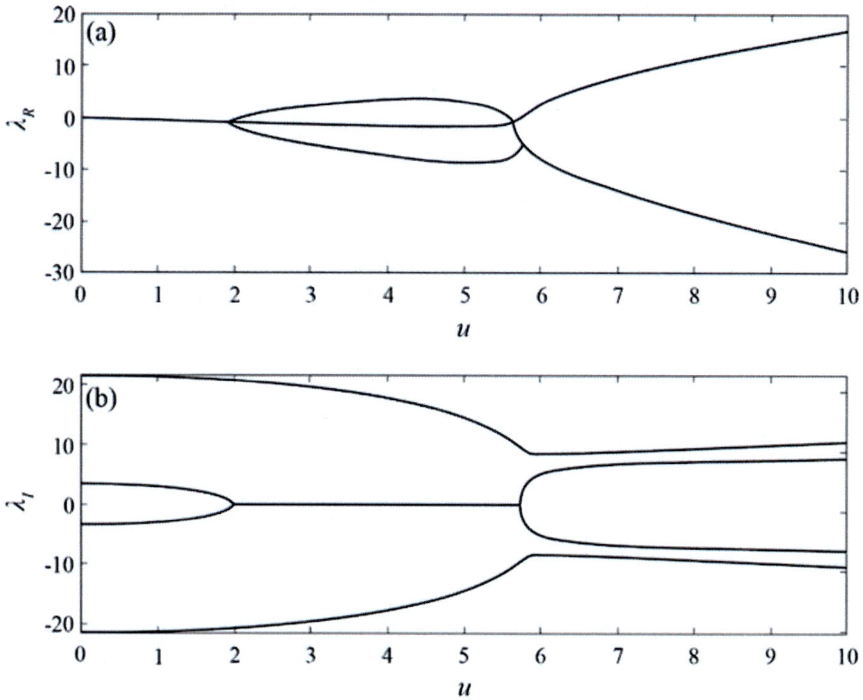

Fig. 8.6 (**a**) Real and (**b**) imaginary parts of the eigenvalues for a cantilevered cylinder in axial flow for $\beta = 0.5$, $\varepsilon = 10$, $\Pi_0 = 0$, $c_N = c_T = 0.1$, $f = 0.8$, $C_b = 0$, $x_e = 0.01$

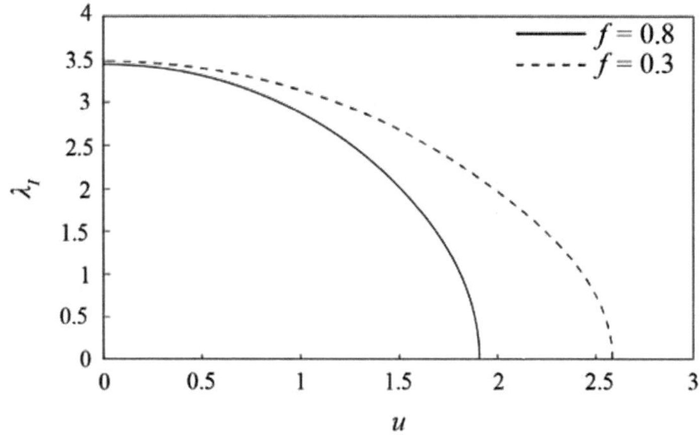

Fig. 8.7 Variation of the system's frequency versus flow velocity for two different values of parameter f: the lower value of f leads to a higher critical flow velocity for buckling

$u = 2.15$ for $c_N = c_T = 0.2$. The parameters c_N and c_T represent the viscous effects on the structure. In the equation of motion, c_T shows up as a tension term, and as its value is increased, the effective tension on the structure increases. An increased tension in the structure results in an increased frequency for any given flow velocity, and again it takes longer for the frequency to go to zero, and as a result the onset of buckling is delayed.

8 Nonlinear Behavior of a Cantilevered Cylinder

The linear analysis of a cylinder in axial flow shows that both for a cylinder supported at both ends and a cantilevered cylinder, the structure loses its stability by buckling as the flow velocity is increased. The question then is what will happen after the onset of instability. Intuitively, we know that when a cylinder buckles, it goes to a nonzero amplitude equilibrium. And after our nonlinear analysis of a pipe supported at both ends, we know that the amplitude of the buckled structure increases with increasing flow velocity. But, what will happen after buckling? Especially for a cylinder in axial flow, will there be any dynamic response? Will the cylinder oscillate at any flow velocity or will the buckled response be the only response that we can observe as we increase the flow velocity? To answer these questions, and any other question related to the behavior of a cylinder after the first point of instability, we need a nonlinear model. Here, we consider a nonlinear model for a cantilevered cylinder in axial flow derived by Lopes et al. (2002). This equation in its dimensionless form is:

$$
\begin{aligned}
&\eta'''' + 2C_M u\beta^{\frac{1}{2}}\dot{\eta}' + \frac{1}{2}u\varepsilon c_N\beta^{\frac{1}{2}}\dot{\eta} + \frac{1}{2}u^2\varepsilon c_N\eta' - \left[-C_M u^2 + \frac{1}{2}u^2\varepsilon c_T(1-\xi) - \frac{1}{2}u^2 c_b\right]\eta'' \\
&+ (1 + (C_M - 1)\beta)\ddot{\eta} + \frac{7}{2}C_M u\beta^{\frac{1}{2}}\dot{\eta}'\eta'^2 + C_M u^2\eta'^2\eta'' - \frac{3}{2}C_M\dot{\eta}\eta'\dot{\eta}' - \frac{3}{2}C_M u\beta^{\frac{1}{2}}\dot{\eta}\eta'\eta'' \\
&+ \frac{1}{4}u^2\varepsilon c_N\eta'^3 - \frac{1}{2}u^2\varepsilon c_T(1-\xi)\eta'^2\eta'' - \frac{1}{2}u^2 c_B\eta'^2\eta'' + \frac{1}{4}u^2\varepsilon c_T h\eta'^3 \\
&- \frac{1}{2}u^2\varepsilon c_T h(1-\xi)\eta'^2\eta'' + 4\eta'\eta''\eta''' + \eta''^3 + \eta''''\eta'^2 - \frac{1}{2}u^2\varepsilon c_N\dot{\eta}\int_0^\xi \eta'\dot{\eta}'d\xi \\
&- \frac{1}{4}\varepsilon c_N\beta\eta'\dot{\eta}'^2 - \frac{1}{4}u\varepsilon c_N\beta^{\frac{1}{2}}\eta'^2\dot{\eta}' - \frac{1}{4}\frac{\varepsilon c_N}{u}\beta^{\frac{3}{2}}\dot{\eta}^3 - (1-\beta)\eta''\int_\xi^1\int_0^\xi\left(\dot{\eta}'^2 + \eta'\ddot{\eta}'\right)d\xi d\xi \\
&+ 2C_M\left(\beta\dot{\eta}' + u\beta^{\frac{1}{2}}\eta''\right)\int_0^\xi \eta'\dot{\eta}'d\xi - C_M\eta''\int_\xi^1\left(\beta\ddot{\eta}\eta' + 2u\beta^{\frac{1}{2}}\dot{\eta}'\eta' + u\eta''\eta'\right)d\xi \\
&+ \eta'(1 + (C_M - 1)\beta)\int_0^\xi\left(\dot{\eta}'^2 + \eta'\ddot{\eta}'\right)d\xi + \eta''\int_\xi^1\left(\frac{1}{2}u^2\varepsilon c_b\eta''\eta' + \frac{1}{4}\varepsilon c_T\beta\dot{\eta}^2\right)d\xi \\
&+ \frac{1}{2}u^2\eta''(\varepsilon c_T - \varepsilon c_N)\int_\xi^1\left(\eta'^2 + \frac{\beta^{\frac{1}{2}}}{u}\eta'\dot{\eta}\right)d\xi = 0,
\end{aligned}
\tag{8.59}
$$

in which the dimensionless parameters are the same parameters that we had seen in the linear equation of the cylinder. In the original equation of this nonlinear model, gravity is also considered, but we have dropped the gravity-related terms in the version of the equation that we use here. This equation might seem very long and complicated, but the steps that have been taken in deriving it are the same steps as what we discussed in the derivation of the equation of motion for the linear model, but instead of keeping terms up to order one only, terms up to order three have been kept in the equation, and as a result we see several more terms in the equation. There are of course some other important details in the derivation of this nonlinear model, which we do not discuss here.

To solve this equation, similar to the method that we used in Chap. 7 to solve a nonlinear pipe, we use the Galerkin technique to discretize the PDE and then use a Runge-Kutta solver to solve the set of nonlinear ODEs that are obtained after the discretization of the PDE. Here, we consider a sample set of parameters, and we discuss the observed response of the system at some sample flow velocities. The parameters that we use are $\beta = 0.5$, $\varepsilon = 10$, $\Pi_0 = 0$,$C_n = C_t = 0.05$, $f = 0.8$, $C_b = 0.2$, $x_e = 0.01$. Based on a linear analysis, we find that the system buckles at $u_b = 1.9$—as expected for a cantilevered cylinder in axial flow. At flow velocities before this critical value, any initial disturbance given to the cylinder will damp out to zero and the cylinder goes back to its original equilibrium position. At flow velocities immediately after this critical flow velocity, we expect to observe a cylinder that is buckled in its first mode and this is exactly what we observe in Fig. 8.8a, as an example of a buckled cylinder at $u = 2$. At higher flow velocities, we observe that the cylinder buckles at its second mode, as Fig. 8.8b shows for a sample case of $u = 4$.

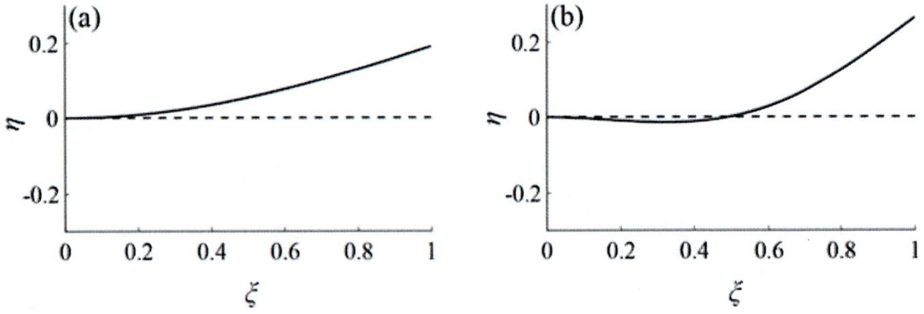

Fig. 8.8 Buckled cantilevered cylinder at (**a**) its first mode at $u = 2$ and (**b**) its second mode at $u = 4$

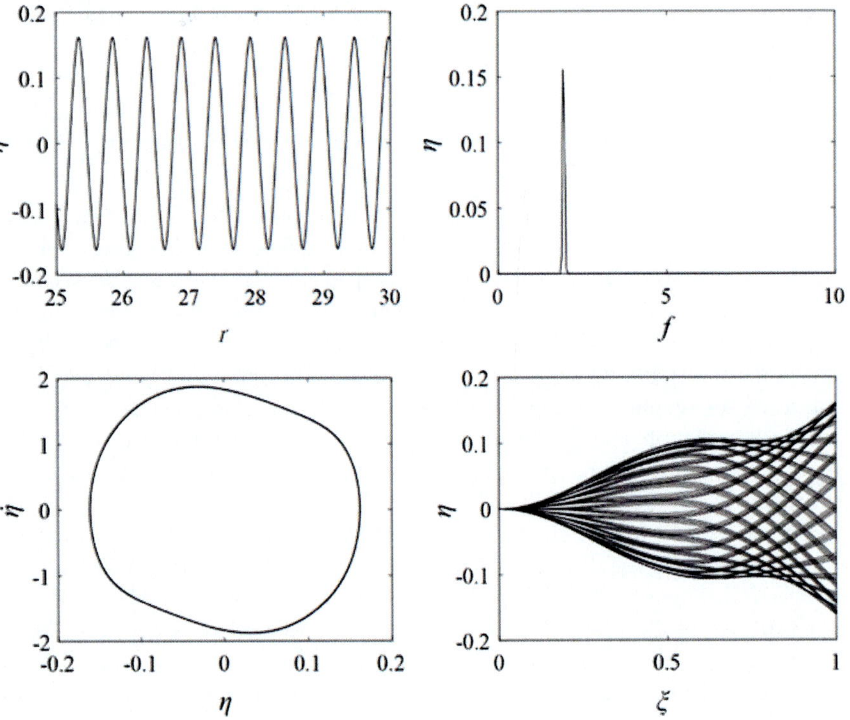

Fig. 8.9 Flutter of a cantilevered cylinder at its second mode at $u = 5.25$

It turns out that at even higher flow velocities, a Hopf bifurcation occurs and the cylinder starts oscillating, mainly in its second mode. Figure 8.9 shows the response of the cylinder at $u = 5.25$, as an example of these second mode oscillations. The time history clearly shows a periodic response and the FFT of the response contains

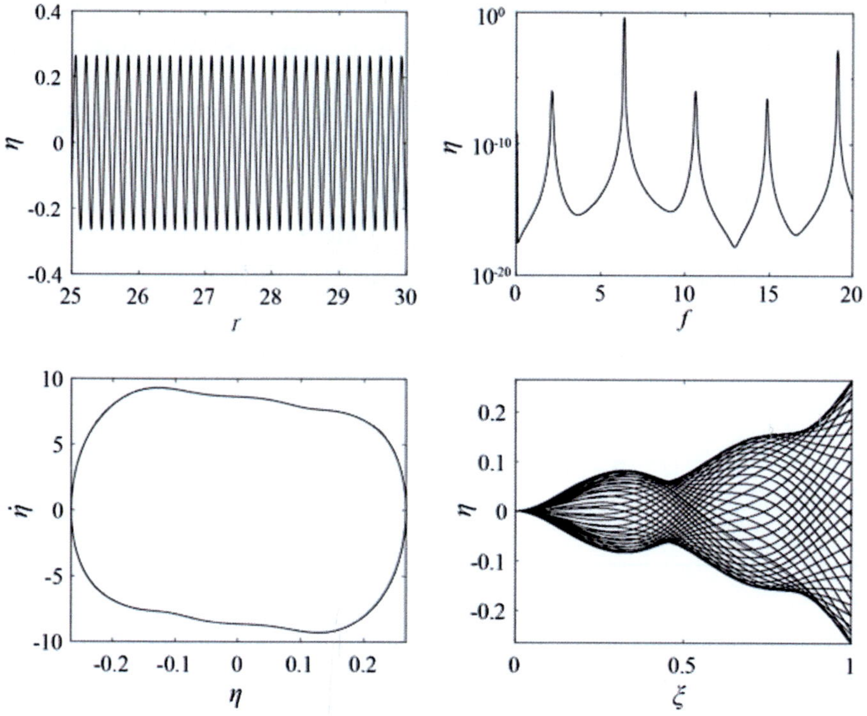

Fig. 8.10 Flutter of a cantilevered cylinder at its third mode at $u = 8.5$

one peak at a dimensionless frequency of ~2. The phase plane is a closed curve, and the snapshots of the cylinder as it oscillates confirm a second mode response, with a localized region of low-amplitude response at $\xi \simeq 0.8$.

At even higher flow velocities, we observe flutter at mainly the third mode of the structure. This is shown for a sample case of $u = 8.5$ in Fig. 8.10. Again, periodic response is obvious in the time history, the FFT shows a peak frequency at a dimensionless frequency of ~6 – clearly a higher frequency than the frequency of oscillations in the previous sample case, which suggests that a higher mode of the structures is excited. The snapshots of the cylinder's oscillations resemble the shape of the third beam mode, with two localized low-amplitude regions around $\xi = 0.4$ and 0.8.

These are only samples of the responses that are observed in a cantilevered cylinder in axial flow. Changing the system parameters changes the details of the observed response. Lopes et al. (2002) in their analysis of this system also observed period doubling in the response, in which a frequency equal to half the fundamental frequency shows up in the FFT plot of the response. Other types of response are also likely to be observed as one changes the system parameters. Here, we do not discuss all possible types of responses that can be observed in this system. The goal here was to show what we can obtain from a nonlinear analysis of the system. Particularly, it is important to know that dynamic responses are also observed in cylinders in axial flow, despite the fact that their first point of instability is always a static one, independent from their boundary conditions. It is also important to note that dynamic responses in the form of periodic, quasiperiodic, and even chaotic oscillations have also been observed for a cylinder supported at both ends and placed in axial flow.

Chapter 9
Vortex-Induced Vibrations of Flexible Beams

1 Introduction

When a flexible beam is placed perpendicular to an incoming flow, it can oscillate, similarly to what a flexibly mounted rigid body does as we discussed in Chaps. 4 and 5. If the cross section of the flexible beam is a bluff cross section, then one can expect to observe VIV and galloping-type responses, similarly to the 1 DOF and 2 DOF systems. If the cross section of the flexible structure is streamlined, then flutter could be observed. Well-known examples for observing such oscillations in flexible structures are oscillations in transmission lines on a windy day. If the cross section of the transmission line is mainly circular, then VIV-type oscillations are observed. If, somehow, the cross section of the transmission line is noncircular and bluff, then galloping-type oscillations can be observed. One of the ways that the cross section of the transmission lines can become asymmetric is when after a snowy or rainy day, the below-freezing temperature results in formation of ice on them and makes their cross sections noncircular. The resulting large-amplitude oscillations are examples of galloping, which could create oscillations with amplitudes of ten (or more) diameters. An example for possible oscillations of a flexible beam with a streamlined cross section is flutter of airplane wings. If the airplane's wing is "flexible enough" and the airplane's speed is "high enough", then the wing can start oscillating in a limit cycle response. Since the wing is a cantilever beam, these oscillations can be observed mainly at the tip of the wing.

In a flexible structure, the flow forces could change along the length of the structure. In the meantime, as we discussed in Chap. 6, the response of a continuous structure can be considered as a summation of many modes of the structure, the amplitude of each of which varies along its length. This makes the problem of a continuous beam in flow a problem that depends both on time and space (the location along the length of the structure): A problem that can be described using PDEs.

In this chapter, we focus on a flexible beam placed in fluid. We first consider the case where the fluid is otherwise still, i.e., it can move only if the beam moves

Y. Modarres-Sadeghi, *Introduction to Fluid-Structure Interactions*,
https://doi.org/10.1007/978-3-030-85884-1_9

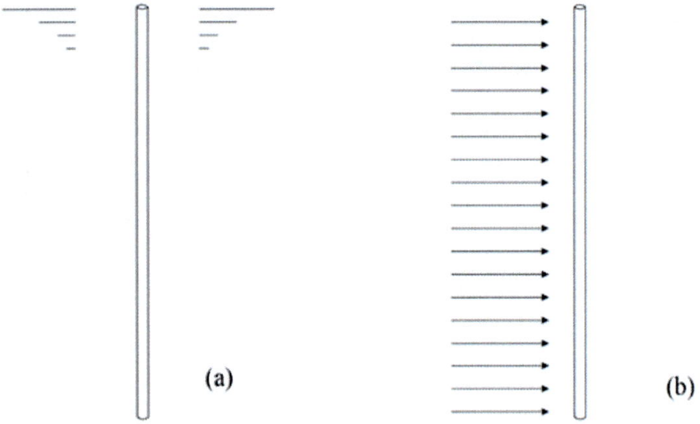

Fig. 9.1 (**a**) A flexible cylinder placed in otherwise still fluid, and (**b**) a flexible cylinder placed in flow perpendicular to its longs axis

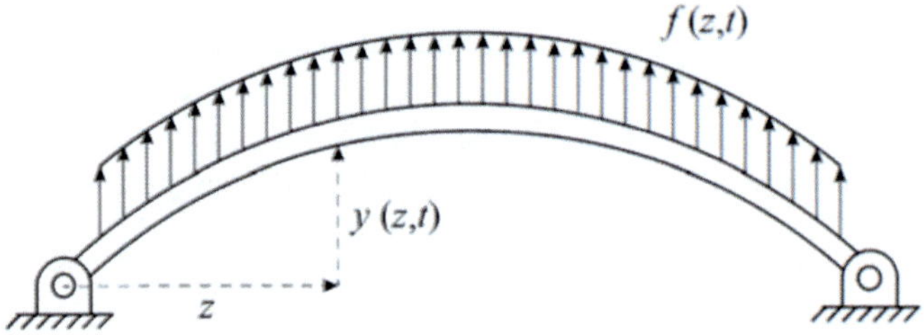

Fig. 9.2 A flexible beam under distributed external force

(Fig. 9.1a). Then we consider the case where the beam is placed in flow, and the flow is perpendicular to the long axis of the structure (Fig. 9.1b). We will consider only the case where the structure is a bluff body with a circular cross section, where VIV is expected to occur. Examples of where VIV can be observed in beams are in risers in offshore oil platforms, and mooring lines in a floating offshore wind turbine.

2 A Beam Placed in Otherwise Still Fluid

As a first step toward understanding the behavior of a beam in flow, we consider a beam in otherwise still fluid (Fig. 9.2). The beam equation, as we discussed in Chap. 6, is

$$\rho A \frac{\partial^2 y}{\partial t^2} + c \frac{\partial y}{\partial t} + EI \frac{\partial^4 y}{\partial z^4} - T \frac{\partial^2 y}{\partial z^2} = f(z, t), \tag{9.1}$$

in which y is the beam's displacement in the direction perpendicular to its long axis, ρ is the beam's density, A its cross-sectional area, c its structural damping, E its modulus of elasticity, I its moment of inertia, T is an external tension that is applied on the beam, and $f(z, t)$ is the external force acting on the beam. When the beam is in contact with fluid, the external force acting on the beam, $f(z, t)$, will be due to the fluid, and to consider the coupled fluid and structure system, we should relate this external force to the motion of the structure. As we discussed before, Navier-Stokes equations are the complete equations which represent the fluid behavior. However, since solving the Navier-Stokes equations in general is not a trivial task, we sometimes come up with other ways to model the flow forces. One way, as we discussed in Chap. 3 for a flexibly mounted rigid cylinder in otherwise still fluid, is to consider two components for the flow forces: a force proportional to the acceleration of the structure (also called the added mass force) and a force proportional to its velocity (the flow-induced damping). This can be written in the form of a Morison-type equation for a beam:

$$f = F_{fluid} = -C_m(z)\rho A\frac{\partial^2 y(z,t)}{\partial t^2} - \frac{1}{2}\rho C_D(z)D\frac{\partial y(z,t)}{\partial t}\left|\frac{\partial y(z,t)}{\partial t}\right|, \tag{9.2}$$

where the first term on the right-hand side is the added mass force and the second term is the flow-induced damping. In these two terms, $C_m(z)$ is the added mass coefficient, $C_D(z)$ is the drag coefficient, ρ is the fluid density, and D is the beam's diameter. For a beam, the added mass coefficient and the drag coefficient can change along its length, and that is why $C_m(z)$ and $C_D(z)$ are written as functions of z.

Equation (9.2) is very similar to Eq. (3.113) of Chap. 3, where we discussed the Morison equation for fluid forces acting on a flexibly mounted rigid cylinder placed in otherwise still fluid. The main difference is that in Eq. (9.2), y is a function of both time and location, and therefore, we have used the partial derivatives.

To obtain the coupled FSI equation, we plug Eq. (9.2) into Eq. (9.1) to get

$$\rho A\frac{\partial^2 y}{\partial t^2} + c\frac{\partial y}{\partial t} + EI\frac{\partial^4 y}{\partial z^4} - T\frac{\partial^2 y}{\partial z^2} = -C_m\rho A\frac{\partial^2 y}{\partial t^2} - \frac{1}{2}\rho C_D D\frac{\partial y}{\partial t}\left|\frac{\partial y}{\partial t}\right|, \tag{9.3}$$

and since both terms on the right-hand side are functions of the cylinder's displacement, they can be moved to the left-hand side:

$$(\rho A + C_m\rho A)\frac{\partial^2 y}{\partial t^2} + \left(c + \frac{1}{2}\rho C_D D\left|\frac{\partial y}{\partial t}\right|\right)\frac{\partial y}{\partial t} + EI\frac{\partial^4 y}{\partial z^4} - T\frac{\partial^2 y}{\partial z^2} = 0. \tag{9.4}$$

Note that Eq. (9.4) has the same form as the beam equation, and since there is no external force in it (the right-hand side is zero), the steady-state response will be zero and the solution will include the transient response only. This implies that an initial disturbance of the system will result in oscillations that decay to zero.

The beam that is described by Eq. (9.4) has natural frequencies and damped natural frequencies which are different from a beam in vacuum. As we saw in Chap. 6, the natural frequency of a beam in vacuum, with simply supported boundaries for example (Eq. (6.105)), can be written as:

$$f_n = \sqrt{\frac{n^4\pi^2}{4L^4}\frac{EI}{\rho A} + \frac{n^2}{4L^2}\frac{T}{\rho A}}. \tag{9.5}$$

If the same beam is placed in otherwise still fluid, thus following Eq. (9.4), its natural frequencies will include the added mass, and will be written as:

$$f_n = \sqrt{\frac{n^4\pi^2}{4L^4}\frac{EI}{\rho A + C_m\rho A} + \frac{n^2}{4L^2}\frac{T}{\rho A + C_m\rho A}}, \tag{9.6}$$

or if we define the total mass as $M = \rho A + C_m\rho A$, then the natural frequencies will be written as

$$f_n = \sqrt{\frac{n^4\pi^2}{4L^4}\frac{EI}{M} + \frac{n^2}{4L^2}\frac{T}{M}}. \tag{9.7}$$

The added mass coefficient for a fluid around a decaying beam is always positive. We discussed possible negative values for the added mass coefficients in the VIV response of a flexibly mounted rigid cylinder in Chap. 4. However, for a beam in otherwise still fluid, the added mass is always positive, since the structure is always accelerating the fluid away from itself. A positive C_m means that M is always larger than ρA and therefore the natural frequencies of a beam in otherwise still fluid are always smaller than its natural frequencies in vacuum.

Another point that we should note from Eq. (9.7) is that in order to check if a beam placed in otherwise still fluid is tension-dominated or not, we should calculate the ratio of the flexural-rigidity-dominated term and the tension-dominated term in the natural frequency:

$$R = \frac{\frac{n^4\pi^2}{4L^4}\frac{EI}{M}}{\frac{n^2}{4L^2}\frac{T}{M}} = \frac{n^2\pi^2 EI}{L^2 T}, \tag{9.8}$$

which is the same as that for a beam in vacuum as we discussed in Chap. 6, since M appears in both terms and is canceled out in the ratio of the two terms. If $R \gg 1$ then the structure is a flexural-rigidity-dominated beam, and if $R \ll 1$, then the structure is a tension-dominated beam or a string.

3 A Beam with a Circular Cross Section Placed in Flow

3.1 General Considerations

If a flexible beam supported at both ends and with a bluff cross section is placed in fluid flow such that its long axis is perpendicular to the direction of the incoming flow, then vortices are shed in the wake of the structure (Fig. 9.3). This is very similar to what we discussed in Chap. 4 for flexibly mounted structures. In the case of a flexible structure, every point along the length of the structure can move in the crossflow and inline directions (the y- and x-directions, respectively, in Fig. 9.3), and therefore there is no pure CF or pure IL oscillations. If we consider a beam with a circular cross section, every point on the beam acts similarly to a 2 DOF flexibly mounted cylinder. The vortices that are shed from each section will result in fluctuating forces with a frequency of f_s in the CF direction and $2f_s$ in the IL direction. As the flow velocity is increased from zero, the shedding frequency increases following the Strouhal law until it matches one of the natural frequencies of the structure in the CF direction, e.g., the natural frequency of mode n (Fig. 9.4). In that case, lock-in can occur in the CF direction, i.e., the cylinder will oscillate in the CF direction with a frequency equal to the shedding frequency, f_s. At the same reduced velocity, the shedding in the IL direction produces forces with a frequency of $2f_s$. Then for the structure to be able to lock-in in the IL direction as well, it needs to have a natural frequency at $2f_s$. As we discussed in Chap. 6, the natural frequencies of a simply supported tension-dominated beam in vacuum increase linearly with the mode number. This is also true for simply supported tension-dominated beams placed in otherwise still fluid, according to Eq. (9.7). Besides, the natural frequencies of the beam in the CF and IL directions are the same, since the beam has a symmetric

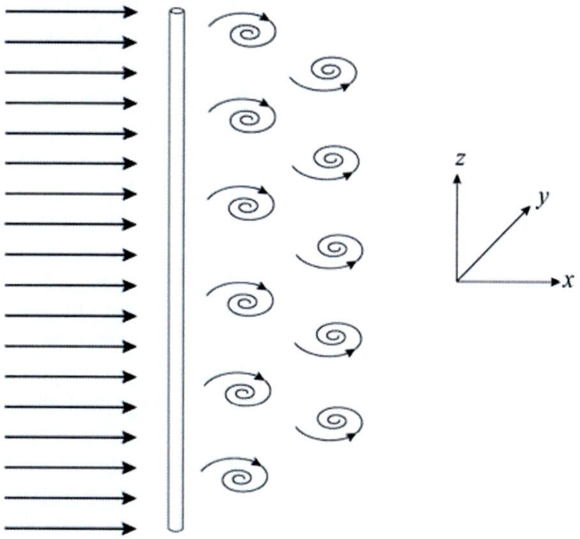

Fig. 9.3 Schematic of a flexible cylinder placed in flow perpendicular to its long axis, and the vortices that are shed in its wake

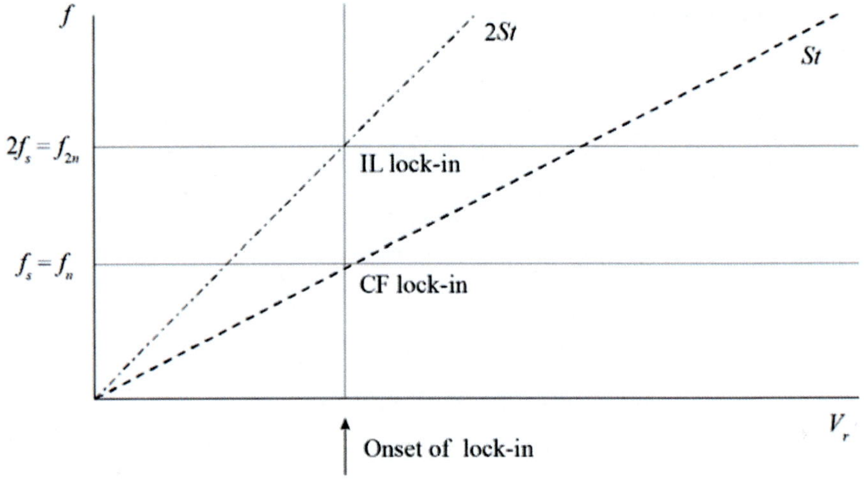

Fig. 9.4 Excited frequencies in the CF and IL directions at a given reduced velocity: f_n is the natural frequency of mode n that is excited in the CF direction and $2f_n$ is the natural frequency of the mode that is excited in the IL direction

cross section. Therefore, if mode n is excited at a frequency of f_s in the CF direction, then there must be a mode $2n$ at a natural frequency of $2f_s$ in the IL direction, which can be excited at the same time. If this occurs, we then observe lock-in in both the CF and IL directions at the same time.

We have to note that since flow forces and possibly even flow velocity can change along the length of a flexible beam, it is very likely that lock-in occurs over a portion of the length of the beam, while the rest of the beam does not lock-in. In this case, oscillations start at the lock-in location and travel along the length of the beam. Then the energy goes into the beam from the flow in the lock-in range and is damped at other places along the length of the beam, as the wave travels. Figure 9.5 is an example of a beam placed in flow with a linearly variable flow velocity. Since the flow velocity changes linearly along the length of the beam, the shedding frequency also changes linearly, following the Strouhal law:

$$f_s = \frac{StU}{D}. \tag{9.9}$$

Then the linearly changing shedding frequency could match the natural frequencies of different modes of the beam at different locations. This is shown in the figure for two modes p and q. Therefore, lock-in can occur at more than one location along the length of the beam, and the beam can be excited at more than one location. As a result, there will be more than one traveling wave that are initiated at these locations and move along the length.

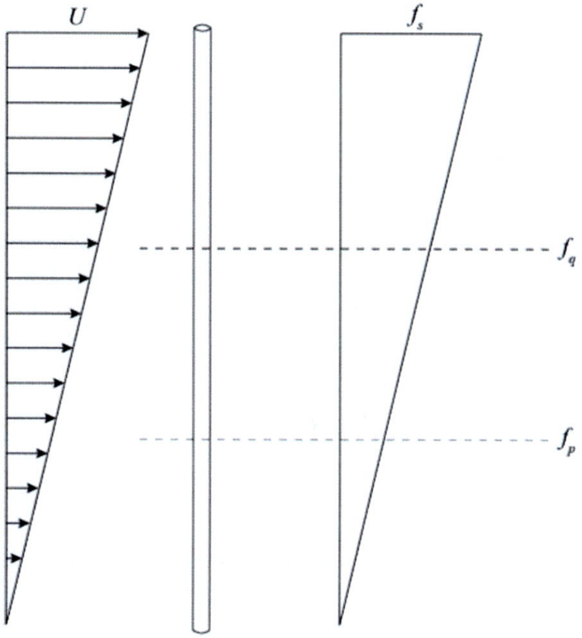

Fig. 9.5 Local lock-in along the length of a beam under a linearly sheared incoming flow. Two different modes of the beam, modes p and q, can be synchronized with the shedding frequencies at two different locations along the length of the beam

The dimensionless parameter that we use as the independent variable in discussing VIV of flexible beams is the reduced velocity, the same parameter that we used for VIV of a flexibly mounted structure in Chap. 4:

$$V_r = \frac{U}{f_n D}, \tag{9.10}$$

in which f_n is the natural frequency of the beam's first mode in otherwise still fluid. Even if higher modes are excited, we still use the natural frequency of the first beam mode in calculating the reduced velocity, so that we can distinguish between the reduced velocity at which the first mode is excited and the reduced velocity at which higher modes are excited. Also, to show the frequency of oscillations, we use the normalized frequency, f^*, which is defined as the frequency of oscillations divided by f_n.

3.2 The Form of Equations of Motion

The equations of motion that represent VIV of a flexible beam consist of two beam equations, one in the CF and one the IL direction (since a beam undergoing VIV can oscillate in both directions) subjected to flow forces in those two directions:

$$\begin{aligned} \rho A\ddot{y} + c\dot{y} + EIy'''' - Ty'' &= f_y(x, \dot{x}, \ddot{x}, y, \dot{y}, \ddot{y}, z, t), \\ \rho A\ddot{x} + c\dot{x} + EIx'''' - Tx'' &= f_x(x, \dot{x}, \ddot{x}, y, \dot{y}, \ddot{y}, z, t). \end{aligned} \tag{9.11}$$

The left-hand sides of these equations represent a linear beam equation, and therefore they are identical, with the only difference being that the first equation is in terms of y (oscillations in the CF direction) and the second equation is in terms of x (oscillations in the IL direction). The right-hand sides are the fluctuating flow forces in the crossflow, f_y, and inline, f_x, directions, respectively. The two equations are decoupled on the left-hand side: The left-hand side of the first one is a function of y and the left-hand side of the second one is a function of x. The forces on the right-hand side, however, are functions of displacements, velocities, and accelerations in both the IL and CF directions, and it is through these forces that the two equations are coupled.

As we discussed in Chap. 4, these fluctuating forces are a combination of the fluctuating lift, f_L, and fluctuating drag, f_D, forces. The difference is that the fluctuating lift and drag are in the directions, respectively, perpendicular to and in line with the relative velocity, while f_y and f_x are in the directions perpendicular to and in line with the incoming flow. These directions (x and y) do not change as the cylinder oscillates, but the directions of lift and drag do. For small-amplitude oscillations (and VIV could be considered to be small-amplitude oscillations), the fluctuating crossflow force is comprised mainly of the fluctuating lift force, and the fluctuating inline force of the fluctuating drag force. Since the fluctuating lift has a frequency equal to the shedding frequency, f_s, and the fluctuating drag has a frequency twice the shedding frequency, $2f_s$, then the fluctuating crossflow will have mainly a frequency of f_s and the fluctuating inline a frequency of $2f_s$.

3.3 A Typical Response

Figure 9.6 shows a typical VIV response of a flexible cylinder. The plots on the first row of this figure show the magnitude of the cylinder response as a function of its length (the z-direction) for both the CF and IL directions, for a cylinder with simply supported boundary conditions. It is clear from the amplitude plot of the CF direction that the second beam mode is excited in this direction (there are two local maxima and a local minimum, which resembles the node in a second mode response). The magnitude plot in the IL direction corresponds to a fourth-mode response. This is clear from the four local maxima and the three local minima, which is what one would expect to observe in the fourth mode shape of a simply supported beam. The amplitudes in both directions are normalized by the diameter of the cylinder—this is the same dimensionless number that we used in Chap. 4 to represent the VIV response of a flexibly mounted cylinder. The magnitudes both in the CF and IL directions are asymmetric with respect to the center of the beam. Although these are the results from a case where the incoming flow velocity was uniform along the length of the cylinder, and the cylinder's boundary conditions were identical at both ends. The observed asymmetry in the response is not surprising, due to the three-dimensionality of flow in the wake of a flexible beam. The wake of a cylinder, even if the cylinder is rigid, is three-dimensional. Flexibility of the structure amplifies this three-dimensionality. Due to the cylinder's flexibility, it bends in the direction of flow as a result of the mean drag that acts on the cylinder, and a static deflection is always observed independent from the observed oscillations. This static deflection also increases the three-dimensional effects in the wake, which in turn makes it more likely for the structure to respond asymmetrically.

The plots on the second row in Fig. 9.6 show the frequency contents of the displacement time histories along the length of the cylinder. The frequency is normalized by the natural frequency of the cylinder in otherwise still fluid. The frequency plots show a dimensionless frequency of ~2.2 in the CF direction and a dimensionless frequency of ~4.4 in the IL direction. This corresponds to a 2:1 ratio

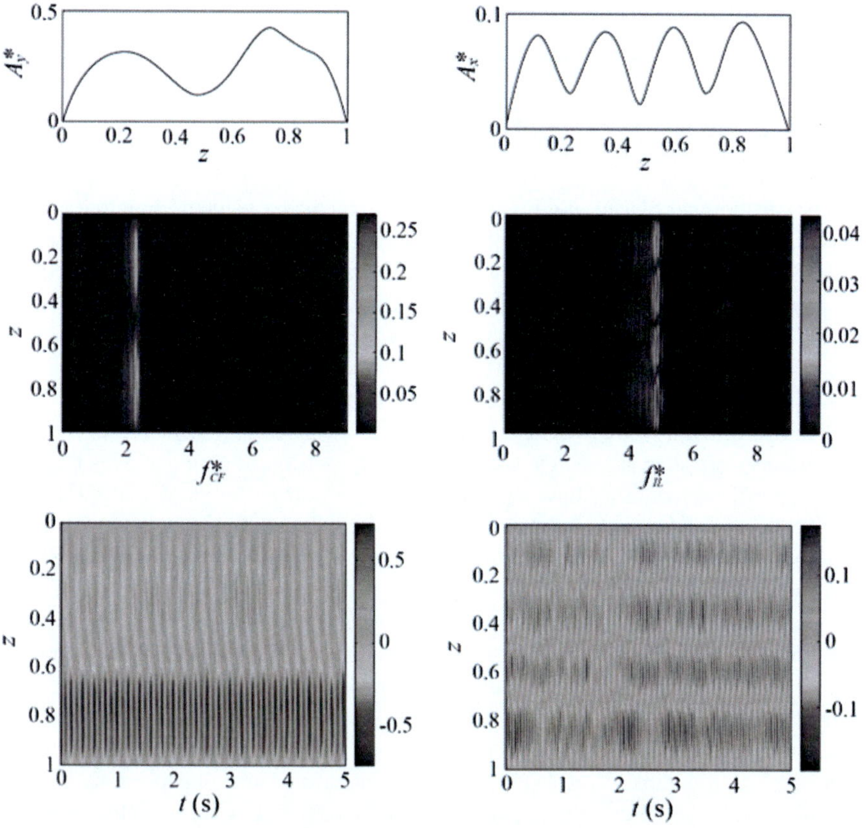

Fig. 9.6 (**a**, **b**) The CF and IL magnitudes of oscillations along the length of a flexible beam undergoing VIV, together with (**c**, **d**) the frequency contents of the CF and IL displacements, and (**e**, **f**) their corresponding time histories along the length of the beam (Seyed-Aghazadeh et al., 2019)

between the frequencies observed in the IL and CF directions. The excited structural modes can also be observed in the frequency plots of the second row of the figure: a 2nd mode in the CF and a 4th mode in the IL direction. This 2:1 frequency ratio is what is expected in the VIV response of a flexible cylinder, similar to what we saw in the VIV response of a 2 DOF flexibly mounted rigid cylinder in Chap. 4.

The plots on the third row of Fig. 9.6 show the time histories of the cylinder response along its length both in the CF and IL directions. The color in these contour plots shows the magnitude of oscillations and the contour plots are made by putting together all time histories of all the points along the length of the cylinder together. The CF plot (the plot on the left in the third row) shows that the peak magnitudes are smaller for the local maximum that is located at the upper part of the plot ($z \sim 0.2$), which is in agreement with what the magnitude plot of the first row shows. The IL plot of the third row also shows the four local maxima in the IL direction that we had also seen in the magnitude plot of the first row. These 3D plots are helpful because they give information on the traveling or standing wave response of the system. It is clear from the CF plot that a peak in the upper half of the cylinder moves downward as time passes.

3.4 The Wake

A flexible cylinder placed in flow deflects due to the mean value of drag that acts on it. This is shown in sample results of Fig. 9.7. This is an example of a relatively long cylinder, in which modes 6 and 12 are excited in the CF and IL directions, respectively. This is clear by looking at the CF and IL time histories of the system in Fig. 9.8. Six local peaks are observed in the CF direction and twelve peaks in the IL direction. In both the CF and IL directions, traveling waves are observed that travel from lower values of z to higher values in the plots of Fig. 9.8.

An overall view of the wake of the cylinder is shown in Fig. 9.7a. The plot shows that vortices are shed at an angle with respect to the deflected cylinder. This is due to the nonuniform incoming flow velocity, as shown in the flow profile in the figure. The flow velocity is higher at the lower end of the structure and lower at its upper end. This nonuniform incoming flow represents the more realistic cases of flexible structures in contact with flow. It is expected that the flow velocity varies along the length of the structures, because they are long and flexible. Consider a riser in an offshore oil platform that goes thousands of meters into the ocean. Or mooring lines that keep a floating offshore wind turbine in place. These structures experience variable incoming flow velocities along their lengths. This means that, even in the case of a constant diameter, the reduced velocity changes along the length of the structure. If we think of each point of the structure as one independent flexibly mounted cylinder, then they all are subject to different reduced velocities, and therefore different points of the structure can lock in with the flow at different incoming flow velocities. In the example of Fig. 9.7, if the cylinder locks in with the flow at some point along its length, then VIV starts at that point and the structure

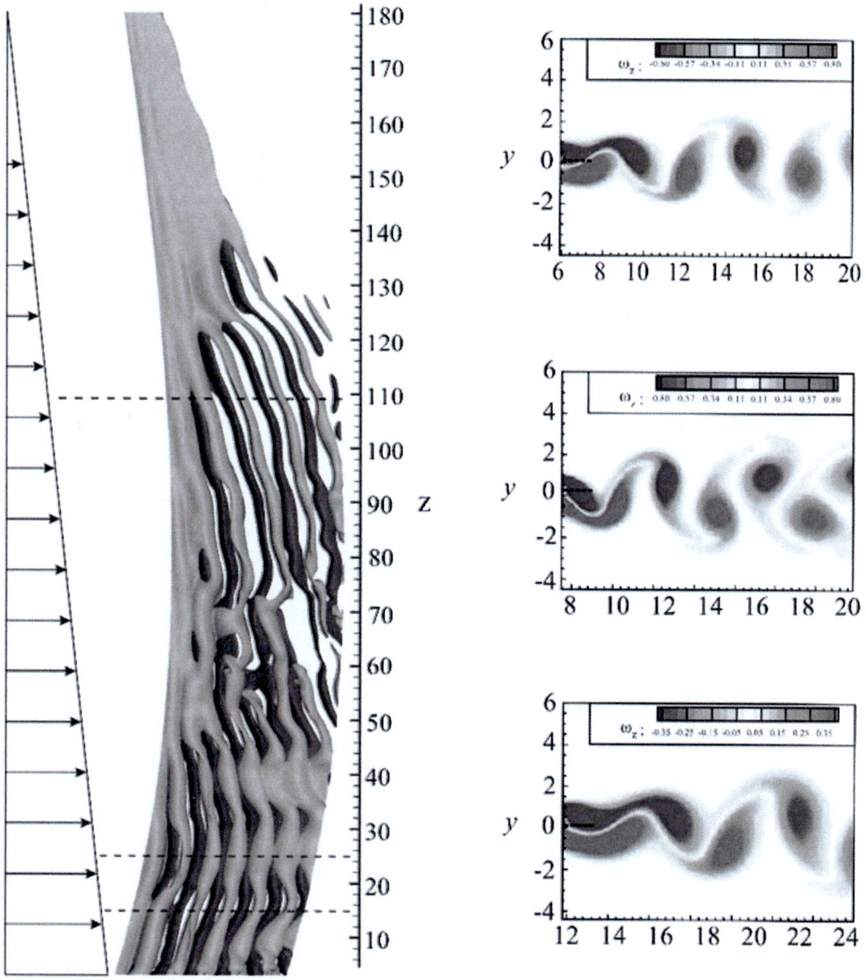

Fig. 9.7 Wake visualization of a flexible cylinder undergoing VIV based on the results from Direct Numerical Simulations (DNS). (Bourguet et al., 2011)

is forced to oscillate in that region. Then, since we have a continuous system, the oscillations travel to the rest of the structure, creating a traveling-wave-type response.

Figure 9.7 shows samples of the wake at different cross sections of the beam. We observe that vortices are shed from the two sides of the cross section and form a wake that resembles the wake of a flexibly mounted cylinder undergoing VIV. While the shedding pattern is similar to the 2S shedding that we observed in Chap. 4 for flexibly mounted cylinders, the wake could get much more complicated at the cross sections of a beam.

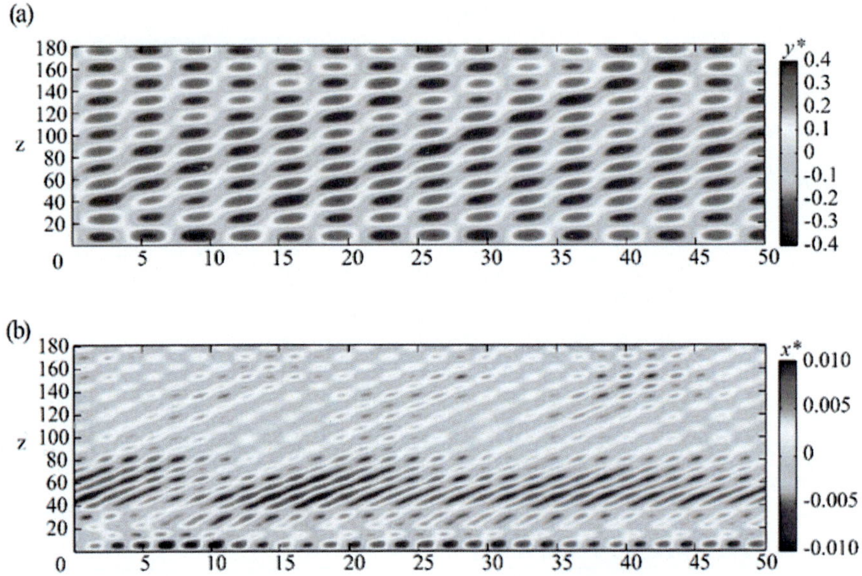

Fig. 9.8 Time histories of the (**a**) crossflow and (**b**) inline displacements along the length of the cylinder. (Bourguet et al., 2011)

3.5 Single-Mode Excitation

Theoretically, for a given beam, if the reduced velocity is increased from zero, the shedding frequency also increases from zero, following the Strouhal law. Then the first mode of the beam is always excited in the CF direction, together with the second mode in the IL direction. This will result in a lock-in region in which the first mode in the CF and the second mode in the IL direction are excited (Fig. 9.9). As the reduced velocity is increased, the lock-in region ends and the cylinder stops oscillating. Then the shedding frequency is increased with increasing reduced velocity until it becomes equal to the beam's second mode's natural frequency. Then the second mode in the CF is excited, while the fourth mode in the IL is excited. This will create a second lock-in region with the second CF mode and the fourth IL mode. And this scenario is repeated as the reduced velocity is increased: The lock-in with the second CF mode and the fourth IL mode will end and then the third CF mode and the sixth IL mode lock-in is observed and so on. The scenario in which each mode is excited after the lock-in range of the previous mode is ended assumes that the natural frequencies of the beam are far enough from each other that one natural frequency has enough space (in the reduced velocity axis) to go through a lock-in range and come out of it before the next natural frequency is excited. However, in several cases, this is not possible. The beam's natural frequencies are, in many cases, close to each other, especially when the beam is placed in water, and the added mass effects are more significant.

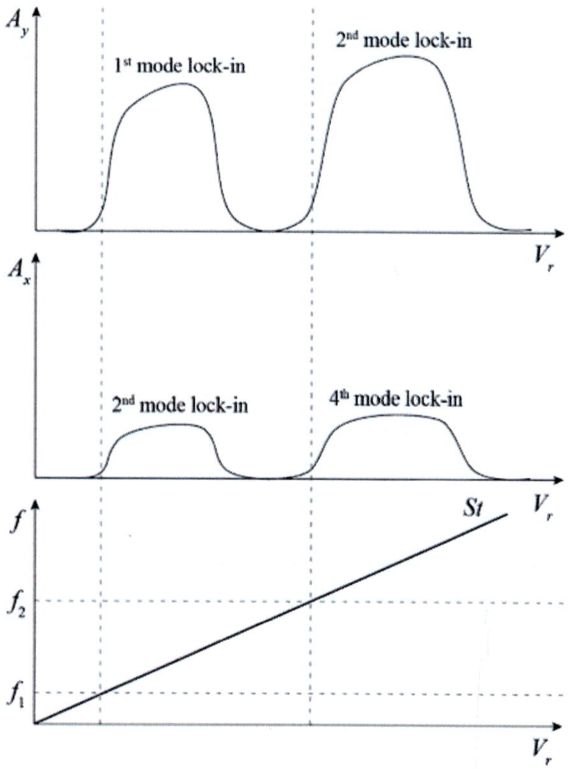

Fig. 9.9 Single-mode excitation of a flexible cylinder in crossflow

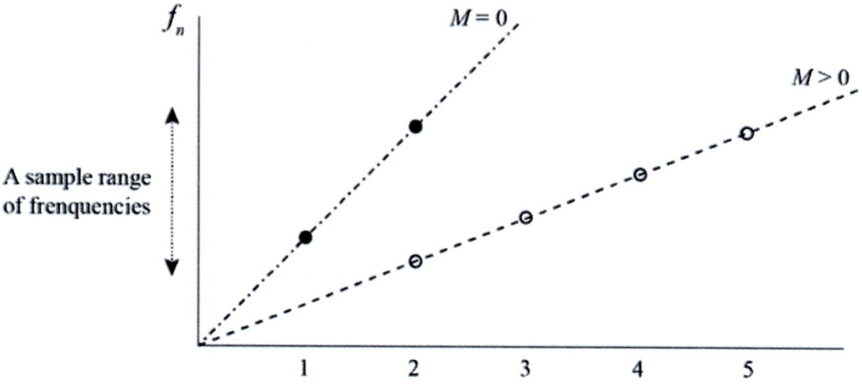

Fig. 9.10 A schematic of the influence of the added mass on the distribution of the natural modes. With no added mass ($M = 0$, in vacuum), only 2 modes (modes 1 and 2) are observed in an arbitrary range of frequencies. In the same range of frequencies, for $M > 0$, 4 modes (modes 2–5) exist

Figure 9.10 shows a schematic of the influence of the added mass on distribution of the natural frequencies for a simply supported tension-dominated beam. Since the added mass is positive in this case, the natural frequency decreases for all modes when the beam is placed in fluid. While still a linear relation exists between the natural frequency and the mode number, in the same range of frequencies (the *y*-axis), more natural modes exist when $M > 0$. This proximity results in cases where while one mode is already excited and is within its own lock-in range, the next mode can be excited too, which results in multi-modal excitation at certain reduced velocities, as we discuss in the following section.

3.6 Multi-Modal Excitation

Several scenarios are possible in which more than one mode could be excited at the same reduced velocity. In one scenario, the first CF and the second IL modes are excited initially. Then, at a higher reduced velocity within this lock-in range, the second CF mode and the fourth IL mode are excited too, resulting in a range of multi-modal excitation. This could continue at higher reduced velocities with multi-modal excitations between higher modes. For example, at some higher reduced velocity, where the second CF mode and the fourth IL mode are still within their lock-in range, the third CF mode and the sixth IL mode can get excited, leading to another multi-modal excitation region, and so on. In this scenario, which is very likely, the oscillations never stop after they start: the observed oscillations will be the result of excitations of different modes (Fig. 9.11).

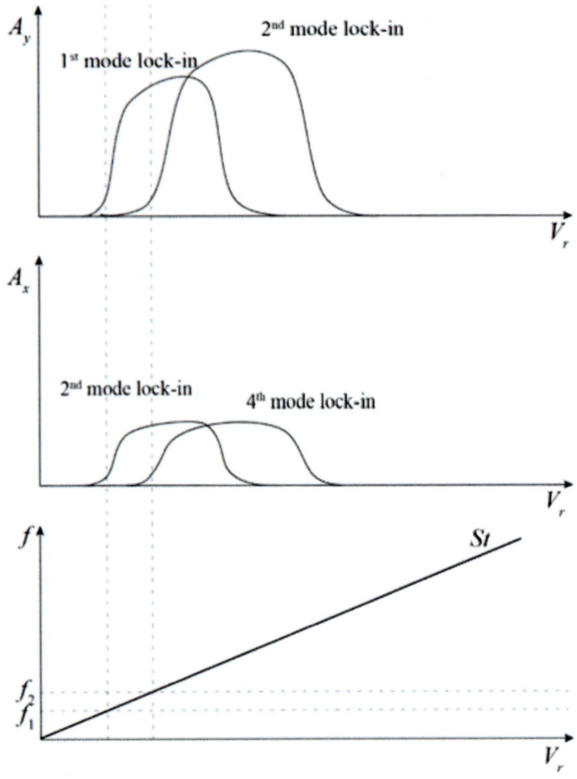

Fig. 9.11 Multi-modal excitation of a flexible cylinder in crossflow

Since in several cases where VIV is observed, the flow velocity (and as a result, the reduced velocity) does not start at zero, but at a nonzero value, it is very likely to observe VIV for a higher mode, n, in the CF and mode $2n$ in the IL direction. We do not necessarily always observe the first mode excitation first and the second mode excitation second, because we do not necessarily increase the flow velocity from zero. We might have a case where, for example, modes 7 in the CF and 14 in the IL direction are excited without any previous modes ever excited, most likely because of the size and the flexibility of the beam and the range of flow velocities.

References

Bearman, P. W. (1984). Vortex shedding from oscillating bluff bodies. *Annual Review of Fluid Mechanics, 16*, 195–222.

Blevins, R. D. (1990). *Flow-induced vibration* (2nd ed.). Kreiger Publishing Company, Malabar, Florida.

Blevins, R. D. (2001). *Flow-induced vibration* (2nd ed.). Krieger Pub.

Bourguet, R., Karniadakis, G., & Triantafyllou, M. (2011). Vortex-induced vibrations of a long flexible cylinder in shear flow. *Journal of Fluid Mechanics, 677*, 342–382.

Carlson, D.W., Currier, T., & Modarres-Sadeghi, Y. (2021). Flow-Induced Vibrations of a Square Prism Free to Oscillate in the Crossflow and Inline Directions. *Journal of Fluid Mechanics, 919*, A2

Copeland, G. S., & Moon, F. C. (1992). Chaotic flow-induced vibration of a flexible tube with end mass. *Journal of Fluids and Structures, 6*, 705–718; also in *American Society of Mechanical Engineers* AMD-Vol. I52, pp. 63–77.

Dahl, J. M., Hover, F. S., & Triantafyllou, M. S. (2006). Two-degree-of-freedom vortex- induced vibrations using a force assisted apparatus. *Journal of Fluids and Structures, 22*, 807–818.

Dahl, J. M., Hover, F. S., & Triantafyllou, M. S. (2007). Resonant vibrations of bluff bodies cause multi vortex shedding and high frequency forces. *Physical Review Letters, 99*, 144503.

Dahl, J. M., Hover, F. S., Triantafyllou, M. S., & Oakley, O. H. (2010). Dual resonance in vortex-induced vibrations at subcritical and supercritical Reynolds numbers. *Journal of Fluid Mechanics, 643*, 395–424.

Den Hartog, J. P. (1956). *Mechanical vibrations* (4th ed.). McGraw-Hill.

Facchinetti, M. L., de Langre, E., & Biolley, F. (2004). Coupling of structure and wake oscillators in vortex-induced vibrations. *Journal of Fluids and Structures, 19*, 123–140.

Fox, R. W., & McDonald, A. T. (1998). *Introduction to fluid mechanics*. Wiley.

Gerrard, J. H. (1966). The mechanics of the formation region of vortices behind bluff bodies. *Journal of Fluid Mechanics, 25*, 401–413.

Gopalkrishnan, R. (1993, February). *Vortex-induced forces on oscillating bluff cylinders*. Doctor of science in oceanographic engineering, Massachusetts Institute of Technology and Woods Hole Oceanographic Institution.

Gurian, T. D., Currier, T., & Modarres-Sadeghi, Y. (2019). Flow forces and a new wake mode in inline vortex-induced vibration of a flexibly-mounted rigid cylinder. *Physical Review Fluids, 4*, 034701.

Holmes, P. J. (1978). Pipes supported at both ends cannot flutter. *Journal of Applied Mechanics, 45*, 619–622.

Inman, D. J. (2014). *Engineering vibrations* (4th ed.). Pearson.

Y. Modarres-Sadeghi, *Introduction to Fluid-Structure Interactions*,
https://doi.org/10.1007/978-3-030-85884-1

Jauvtis, N., & Williamson, C. H. K. (2004). The effect of two degrees of freedom on vortex-induced vibration at low mass and damping. *Journal of Fluid Mechanics, 509*, 23–62.

Lienhard, J. H. (1966). *Synopsis of lift, drag and vortex frequency data for rigid circular cylinders* (College of Engineering Research Division Bulletin 300). Washington State University.

Lighthill, M. J. (1960). Note on the swimming of slender fish. *Journal of Fluid Mechanics, 9*, 305–317.

Lopes, J. L., Païdoussis, M. P., & Semler, C. (2002). Linear and nonlinear dynamics of cantilevered cylinders in axial flow. Part 2: the equation of motion. *Journal of Fluids and Structures, 16*, 715–737.

Modarres-Sadeghi, Y., & Païdoussis, M. P. (2009). Nonlinear dynamics of extensible fluid-conveying pipes, supported at both ends. *Journal of Fluids and Structures, 25*, 535–543.

Modarres-Sadeghi, Y., & Païdoussis, M. P. (2013). Chaotic oscillations of long pipes conveying fluid in the presence of a large end-mass. *Computers and Structures, 122*, 192–201.

Modarres-Sadeghi, Y., Semler, C., Wadham-Gagnon, M., & Païdoussis, M. P. (2007). Dynamics of cantilevered pipes conveying fluid. Part 3: Three-dimensional dynamics in the presence of an end-mass. *Journal of Fluids and Structures, 23*, 589–603.

Morison, J. R., O'Brien, M. P., Johnson, J. W., & Schaaf, S. A. (1950). The forces exerted by surface waves on piles. *Journal of Petroleum Technology, Petroleum Transactions, AIME, 189*, 149–154.

Païdoussis, M. P. (1973). Dynamics of cylindrical structures subjected to axial flow. *Journal of Sound and Vibration, 29*, 365–385.

Païdoussis, M. P. (2014). *Fluid-structure interactions: Slender structures and axial flow, Vol. 1* (2nd ed.). Academic Press.

Païdoussis, M. P. (2016). *Fluid-structure interactions: Slender structures and axial flow, Vol. 2* (2nd ed.). Academic Press.

Païdoussis, M. P., Price, S. J., & de Langre, E. (2010). *Fluid-structure interactions: Cross-flow-induced instabilities*. Cambridge University Press.

Parkinson, G. V., & Smith, J. D. (1964). The square prism as an Aeroelastic nonlinear oscillator. *The Quarterly Journal of Mechanics and Applied Mathematics, 17*, 225–239.

Rao, S. S. (2011). *Mechanical vibrations* (5th ed.). Pearson/Prentice Hall.

Sarpkaya, T. (2004). A critical review of the intrinsic nature of vortex-induced vibrations. *Journal of Fluids and Structures, 19*, 389–447.

Semler, C., Li, G. X., & Paıdoussis, M. P. (1994). The non-linear equations of motion of pipes conveying fluid. *Journal of Sound and Vibration, 169*, 577–599.

Seyed-Aghazadeh, B., Edraki, M., & Modarres-Sadeghi, Y. (2019). The effect of boundary conditions on vortex-induced vibration of a fully-submerged flexible cylinder. *Experiments in Fluids, 60*, 38.

Smith, J. D. (1962). *An experimental study of the aeroelastic instability of rectangular cylinders*. University of British Columbia.

Strogatz, S. H. (2015). *Nonlinear dynamics and chaos: With applications to physics, biology, chemistry, and engineering* (Studies in nonlinearity) (2nd ed.). Westview Press.

Sumer, B. M., & Fredsoe, J. (2006). *Hydrodynamics around cylindrical structures* (Revised ed.). World Scientific Publishing Company.

Thomson, W. T., & Dahleh, M. D. (1998). *Theory of vibrations with applications* (5th ed.). Prentice Hall.

Williamson, C. H. K., & Govardhan, R. (2004). Vortex-induced vibrations. *Annual Review of Fluid Mechanics, 36*, 413–455.

Zdravkovich, M. M. (1997a). *Flow around circular cylinders: Volume 1: Fundamentals*. Oxford University Press.

Zdravkovich, M. M. (1997b). *Flow around circular cylinders: Volume 2: Applications*. Oxford University Press.

Index

Y. Modarres-Sadeghi, *Introduction to Fluid-Structure Interactions*,
https://doi.org/10.1007/978-3-030-85884-1